ETIOLOGY

ETIOLOGY

How to Detect Disease in Your Energy Field Before It Manifests in Your Body

Christina L. Ross, PhD, BCPP

DISCLAIMER

Medical knowledge is constantly changing. As new information becomes available, changes in treatment procedures, equipment and the use of drugs becomes necessary. The author and the publisher have, as far as possible, taken care to ensure that the information given in this text is accurate and up to date. However, readers are strongly advised to confirm that the information, especially with regard to treatments and drug usage, complies with the latest legislation and standards of practice. None of the material in this book is intended to provide a recommendation for diagnosis or treatment of disease. The reader who suspects he or she has a medical problem should see a licensed health care practitioner.

To order additional copies of this book, contact:
Xlibris Corporation
1-888-795-4274
www.Xlibris.com
Orders@Xlibris.com
105195

Acknowledgments

I wish to thank the book's illustrator, René Harrell, for her invaluable creativity and hard work in drawing the illustrations for this book.

It is with deep love and appreciation that I thank my husband, Cameron Ross, for his unending patience and support.

Table of Contents

PREFACE ix

FORWARD xi

CHAPTER ONE 1
What is Disease?

CHAPTER TWO 5
Why Do We Get Sick?

CHAPTER THREE 19
Is There a Way to Prevent Illness?

CHAPTER FOUR 29
How Can We See Disease Coming?

CHAPTER FIVE 35
Is it my Fault I Got Sick?

CHAPTER SIX 41
If I Get Sick What Are My Options?

CHAPTER SEVEN..57
What is Energy Medicine?

CHAPTER EIGHT 73
The Human Biofield

CHAPTER NINE...81
The Five Elements and How they Relate to Health and Healing

CHAPTER TEN..89
Before Conception to Birth

CHAPTER ELEVEN 115
Cell Communication and Disease

CHAPTER TWELVE 125
Morphogenesis and Holography

CHAPTER THIRTEEN 139
How does my Physical Health affect my Emotional Health and Vice Versa?

CHAPTER FOURTEEN 147
How Do My Thoughts and Emotions Affect My Health?

CHAPTER FIFTEEN 153
How does Spiritual health effect psychological health and vice versa?

CHAPTER SIXTEEN 161
How Eating Habits Affect Health

CHAPTER SEVENTEEN 173
The use of both Conventional and CAM modalities to heal

CHAPTER EIGHTEEN 193
How Do I know which CAM is best for my condition?

CHAPTER NINETEEN..209
Living a Poor Quality of Life vs. Dying

CHAPTER TWENTY 217
How Can Disease Change My Life For the Better?

CHAPTER TWENTY ONE 221
Conclusion

CHAPTER TWENTY TWO 223
Resources

GLOSSARY OF TERMS 231

INDEX.. 241

Preface

This book presents a different perspective of disease. Its intent is to provide a resource for readers who want to take an interactive role in their own health and healing process. Disease presents with it an opportunity to change lives for the better. Disease is an opportunity for change not only within an individual but also within the health care system as a whole. This change can occur with advancements in psychological, emotional, spiritual, as well as physiological therapies. As the biotech industry becomes more advanced, molecular biologists are seeing the need to study more than just the physiology of the body—they are beginning to understand the *energy* of the body. Biologists study organisms and physicists study energy. Biophysicists study the human organism not as the sum of its cellular and molecular parts, but as an integrated, sophisticated system of cell communication and information exchange. Our mind and body are in constant communication. Our emotions affect our physiology. Not only does our food fuel us physically, it feeds our minds, our emotions, and ultimately our spirits. Our spirit communicates through our emotions, and when our soul is not feeling well, our emotions can make us sick. Our thoughts can make us physically ill.

This book is a guide to discovering how the global communication system within the human body works. The food we eat, the lifestyle we live, the thoughts we think, and the emotions we feel affect how we manifest disease. It is a resource for understanding how we function as complete human beings. It offers a biophysicist and energy medicine practitioner's perspective of promoting optimal health and options for treatment if we become ill. The reader is encouraged to become empowered by this information, to discuss it with their health care professional, and to set up a health care plan that optimizes their quality of life. Disease is the imbalance of the entire human being—mind, body, and spirit. If we are alert, if we pay attention, we are empowered to not only detect, but also intercept disease before it manifests in our body.

Forward

Etiology (e" te-ol' o-je) [Gr. *aitia* cause + -logy]. The sum of knowledge concerned with causes. In medicine: the study or theory of the origin or cause of any disease or abnormal condition.

We all want to know about the causes of things and often depend on science to provide reliable answers. This is especially so if we find ourselves with a disease or other health issue. Studies of the causes of health problems lie at the core of medicine, and become very personal when we are diagnosed with any medical condition. Often the first question is, "What caused this?" The second question may be, "Why ME?" We might also ask, "What is the most reliable option for treatment?" And, perhaps the most important question is "How could I have avoided this?" The best medicine is the medicine that keeps us healthy and happy and that enriches our spirit.

Our scientific medicine seeks reliable answers to all of these questions through research. However, when we investigate a medical issue, we often find ourselves between two worlds. There is so-called conventional or allopathic medicine, and there is a mix of ancient and modern therapies, often termed complementary, alternative, integrative, or cooperative medicine. Which path should we follow to answer our questions? Ideally, there would be a way to look at the best information from both sides of this topic. The remarkable book you are holding in your hands is a perfect way to begin such a fascinating and often vital look at causes, prevention and treatment.

Dr. Christina Ross has lived in the two worlds of medicine, first as a successful alternative therapist, and then as a successful researcher in the highly skeptical environment of a world-class medical center. From the synthesis of her experiences as a therapist, her biophysical research and a lot of careful study comes this book: a highly readable and integrated perspective on causes, on *etiology*. Her goal: to enable us to prevent health problems before they become serious and to sense when our bodies are slipping out of balance. Plus basic information on the help that is available, and how it works.

Dr. Ross has done laboratory research that will someday save many lives; this book has the potential to save many more from the pain, discomfort and financial burden posed by preventable chronic diseases. It is a book for all of us!

James L. Oschman, Ph.D.
Author of *Energy Medicine: the scientific basis*

To the Reader

This book is written in multidimensions, which means that different people will receive different messages from it depending on their need. If a segment or chapter does not resonate with you, simply turn the page. You do not have to be a biophysicist or a physician to get insight from these chapters. This book is meant to be a resource for people who feel empowered to take control of their own health and healing process.

CHAPTER ONE
What is Disease?

Disease is a language expressed through our body, our mind and our emotions. It is also the language of our soul. Disease not only affects us but those around us, changing lives and life's meaning. Contrary to popular belief, disease does not sneak up on us. There are many signs and signals that disease is imminent, long before we feel symptoms and are diagnosed. These subtle signals, which if ignored, increase in amplitude through signals we call aches and pains. After refusing to pay attention to the acute signs, disease becomes a chronic force in our lives. We have been conditioned to ignore it, silence it with pharmaceuticals, alcohol or pain killers. Disease is designed to alert us that our soul is distressed. Disease stirs up innumerable emotions within us. We have experienced its repercussions, survived its hardest punches, lived with chronic disease that can be debilitating and decrease our quality of life. Incurable disease feels like a death sentence. But what if disease is something other than a dreaded affliction? What if we redefined the word disease? Not just as *dis-ease*, but as an opportunity to understand exactly who we are – as psychological, emotional and spiritual beings. Disease is an opportunity for growth. Disease is information that we are not living our life's purpose. Chronic disease is an opportunity to improve our quality of life. Incurable diseases such as heart disease, cancer, HIV/AIDS, and diabetes are *always* life changing, many times changing our relationships with others, and especially with ourselves.

The concept of disease as information is one I have begun to explore as both a healing practitioner and as a biomedical researcher. The current way we view disease is backward. Physiological disease is the last to manifest. Disease manifests first in our spirit, shifting down into our mind and emotions, finally into our physical body. By the time disease has appeared in our physical body as an adult it has been causing problems spiritually, mentally and emotionally for years. When we feel sick clinicians examine our body, drug it, cut it open and take things out or put things in, perhaps even radiate it. Often the side effects of pharmaceutical treatments deplete our quality of life even further. When we are diagnosed with a disease our first reaction is fear. Fear often turns to anger. Anger can turn to depression. We begin to feel weaker psychologically before we weaken physically. Just thinking about our disease makes us sicker. Our health care providers are compassionate, but unable to empower us, so we begin to feel like victims. Current health care methods are most successful at treating acute illness, but the majority of chronic illness is diagnosed as "etiology unknown", meaning we do not know what causes it, and because there is no cure we must learn to live with it. With incurable illness, we are left to our own devices, often being advised to go home and put our affairs in order - basically ending life as we know it.

As depressing at this is there are ways to feel empowered despite having an incurable disease. There are ways to detect disease long before it manifests as a mild illness much less chronic or incurable. Even with chronic or incurable disease, there are ways of coping with and even reversing the painful symptoms of these diseases. Instead of treating the *symptoms* of disease we can change the morphology of the disease - at its origin, long before it presents physiological symptoms, and before it presents emotional and psychological problems. Disease begins at a much higher level than our physiology, psychology and even our emotions. Disease begins as the result of the depletion of our human spirit. Disease can be detected as unease in our lives, becoming imminent if action is not taken. To treat disease from the physiological level is to overlook where it began. Disease begins as the misalignment of our soul with our life's purpose, followed by frustration initiated by our mind and emotions, with symptoms presented physiologically as the final construct. Because so much time has passed before the physical manifestation of a disease, multiple layers need to be addressed before our physiology can be treated. These layers are not mutually exclusive - an imbalance in one will affect the others. Our spiritual body is more ethereal than our physical body therefore it is not visible to us. Our spiritual body is subtle and mature and does not demand attention like our physical body. Its power comes from acting as a guide to our mental, emotional and physical bodies. It is the space in which all other levels function. Without our spiritual body, we would not exist in physical form. To understand our spiritual body is to understand who we are as an individual. It is from our spiritual self that imbalance and depletion starts - disease begins in our spiritual body.

Disease is instability in our Spirit.

Instability in our soul occurs for many reasons. Unless we are totally in tuned spiritually it will be difficult to know when our spirit becomes unstable. Spiritually aware people are often seasoned yoga practitioners, psychics or those who have fully embraced meditation and enlightenment. Most people are unaware if their soul is happy. Most of us rarely if ever think about our soul. Clerics rarely ever focus on the "soul" during their church sermons. If they do it is usually in terms of life after dead not life before birth. They discuss the soul as if it has human attributes. But we are not human...we are souls having a human experience. Our soul existed before we were born and will continue to exist after we die. Disease is a catalyst by which our soul leaves our body.

Because we do not embrace the spiritual aspect of who we are, we will not know there is instability in our soul until we feel an imbalance in our emotions. Emotions are the bridge between our soul and our mind. For those who are not spiritually aware the only signal that something is amiss is experienced through sadness, depression or boredom. This can occur early in life when we are children or later in life as adults, when we lose our joy or zest for life. Dissatisfaction with life initiates disease. Once this is understood it is important to change how we live our life, or at least change how we react to life and its stressors. Our reaction to stress can either stimulate peace or create more distress in our lives. We can determine how far disease has progressed based on how weak and distressed we are. Being able to detect disease in its early stages is life sustaining and empowering, but it does require introspection. Introspection can be painful, uncomfortable at times. It takes effort to soul search. By living a superficial life we risk our health. We have the ability to be very powerful in the face of disease, preventing it before it manifests and improving the quality of our life despite it. I invite you on a journey to discover who you are - as a mental, physical and a spiritual being. I invite you on a journey towards healing.

CHAPTER TWO
Why Do We Get Sick?

Disease begins long before we feel its effect in our physical body. It starts with our spirit, spreads to our mind and emotions, and ultimately manifests in physical form.

If disease is manifested spiritually long before it becomes physical, how is a baby born with disease?

Disease at any stage of life will affect everyone around the person who has the disease. A baby will incarnate with disease in order to provide its parents and caregivers opportunities for spiritual growth. Some may argue it is unfair that an innocent being should have such a fate, but to that soul it is the greatest act of unconditional love it can express. Many of these beings die before they reach adulthood because their life's work is complete in a short period of time. Blessed are those souls who live selflessly in human form.

Why do adults get sick?

Physically we get sick because our immune system has been compromised. We get sick psychologically and emotionally because we have been traumatized or have a chemical imbalance in our brain. We get sick spiritually because we have forgotten we are divine beings. We incarnate for a purpose. We are here to understand and express unconditional love as well as contribute to the well being of humanity and our planet. When this mission is not being fulfilled we are given feedback from our soul.

Disease can be divided into stages beginning at the spiritual level:

1. *Primary* – our soul is not completing its life's purpose. Disease begins spiritually with the dissociation of mind and emotions from spirit.
2. *Secondary* – psychological and emotional disease is created through traumatic life experiences or physiological and chemicals imbalances in the brain.
3. *Tertiary* – psychological and emotional stages of disease begin to manifest as acute physiological illness – headache, nausea, aches and pains, colds, flu, etc.
4. *Quaternary* – acute phase disease becomes chronic physiological disease – migraines, acid reflux, osteoarthritis, Chronic Obstructive Pulmonary Disease (COPD), irritable bowel syndrome (IBS), etc.
5. *Quinary* – chronic disease becomes incurable – onset of cardiac disease, HIV/AIDs, diabetes, cancer, etc.

In order to understand why we get sick, let's first examine the orthodox medical view of disease - our physiology as separate from the rest of our being. As suggested earlier, disease begins with a compromised immune system. As a population we tend to be exposed to the same type of germs depending where we live. When we move to different parts of the country or the world we often get sick when exposed to different microbes. When this happens our immune system builds up antibodies against these microbes. Acute illness, such as colds, flu and allergies are treatable once our immune system has acclimated to our new environment. Understanding the language of disease can guide us to more nurturing forms of self care such as yoga, meditation, better quality sleep or a better health-sustaining diet. Disease is a sign that change needs to occur. In order to defend ourselves against disease, let's discuss some of the basic causes of physiological disease.

Microbes Cause Disease

Microbes are parasitic microorganisms that prey on the host they invade. There are four types of microbes: bacteria, viruses, fungi and protozoa. Bacteria are single cell organisms that contain DNA, so they can replicate, but their DNA is not in a nucleus like in a human cell. Some bacteria produce spores which can be inactive for a long time, coming to life when our immune system is compromised, wreaking havoc on our bodies. Not all bacteria are harmful. Bacteria in the mouth, stomach and intestinal tract are helpful to digestion and prevention of infection[1]. Most bacteria can be killed with antibiotics, but some have become resistant[2]. Examples of diseases caused by bacteria are cholera, *E. coli*, streptococcus, staphylococcus and salmonella.

Another type of microbe or pathogen is a virus. Viruses contain a core of nucleic acid (RNA or DNA) covered by a layer of protein. They invade our cells and make replicates of themselves. They are inactive outside of human cells, but cannot be killed with antibiotics. Examples of diseases caused by viruses are influenza, herpes, H1N1 (swine flu), measles, chicken pox, HIV/AIDS, small pox and rabies. The third type of microbe is a fungus. Fungi DNA are contained in the nuclei of plants and animals. They have thread-like structures called hyphae that penetrate the organisms they invade. Not all fungi cause diseases, but disease state fungi can usually be contained with topical applications or pharmaceuticals. Diseases caused by fungi include Candida, athlete's foot and ringworm. The fourth type of microbe or pathogen is protozoa. Protozoa are single cell organisms that contain nuclei and a cell membrane similar to human cell structure. Diseases caused by protozoa are malaria, sleeping sickness, diarrhea and Giardia.

Pathogens cause diseases that are very destructive in an immune compromised environment. A healthy immune system is efficient at seeking out and destroying foreign invaders that can make us sick. Our T-cells are programmed to detect if our body is being invaded by a foreign microbe. They know the difference between what is suppose to be in our body and an invader. They do this by differentiating between certain molecules called antigens that are present on the surfaces of all natural cells in our body. These immune cells are programmed to protect us, unlike foreign antigen pathogens or internal disease processes like cancer, which attack us. When a microbe such as bacteria enters our body, white blood cells called macrophages phagocytose or engulf and destroy these intruders. If even one cell in our body gets diseased then certain molecules that are usually hidden from the immune system suddenly appear. T-cells recognize invading microbes and kill the diseased cell. Once the T-lymphocyte recognizes an infected cell, it produces a new set of proteins that it places on the surface of the diseased cell. Those proteins bind to receptors on the infected cell triggering a cascade of events that lead to cell suicide, known as apoptosis[3]. There are trillions of different possible antigens and just as many immune cells produced by our body. In humans there are about ten trillion lymphocytes present at any given time[3]. Our immune systems are capable of detecting and eliminating potential illness long before it takes over our body.

Genetic Defects cause Disease

A genetic disease is due to a faulty gene or group of genes, but not all genetic defects cause disease. Genes are blueprints or instructions for making proteins. Proteins control cell functions, and defects in the blueprints can prevent a cell from functioning properly. Genes are made of deoxyribonucleic acid (DNA), a chemical composed of compounds called nucleotides which are carried on chromosomes within the nucleus of a cell. A gene represents a length of DNA code that contains the nucleotide base sequence needed to make a specific protein. A cell makes a cloned copy of a gene in the form of another type of nucleic acid called ribonucleic acid (RNA). The RNA copy is the molecule that is used to assemble the code into a protein molecule. RNA information can also flow backward as well as forward, to alter the host cell's DNA code[4]. It is important to note that genes cannot turn themselves on and off because they are incapable of activating their own genetic expression. Instead they react to signals from their environment[5].

Genetic Testing

Genetic testing uses the human genome to run a genetic diagnosis of possible risks to inherited disease. Under normal circumstances people carry two copies of every gene – one from their mother and one from their father (except for sex-linked traits which males inherit from their mother). Genetic testing uses biochemical tests to detect the possible presence of genetic disease or mutant forms of genes associated with increased risks of developing genetic disorders. As recently as the late 2000's a human genomic test would cost upwards of $350,000, but recent advances in technology have brought the cost down from between $100 for short sections of the genome, to more than $2,500 for the full genomic test. Genetic testing can diagnose or rule out a specific genetic or chromosomal condition. It can be used to confirm a diagnosis when a particular condition is suspected. Diagnostic testing is not available for all genes or genetic conditions[6].

Genetic testing can also identify one copy of a gene mutation, that when present in two copies, causes a genetic disorder. This type of testing is usually offered to people who have a family history of a genetic disorder and to people in ethnic groups who have increased risk of specific conditions. If both parents are tested, carrier testing can provide information about a couple's risk of having a child with a genetic condition. Prenatal genetic testing can detect fetal genes or chromosomal risks before birth. Genetic testing is also performed on human embryos before in vitro fertilization. Predictive and pre-symptomatic testing detects genetic mutations associated with disorders that appear after birth or later in life, such as Alzheimer's disease and certain types of cancer.

Do All Genetic Mutations Cause Disease?

Along with coding for proteins, genes contain hereditary substances, making genetic diseases inheritable[7], but not predestined. This is because DNA is simply a blueprint not a predetermined fact. Genetic defects can cause disease to develop in a variety of ways. Genetic diseases include cystic fibrosis, Duchenne muscular dystrophy, hemophilia, sickle cell anemia and Marfan syndrome among others[8]. Because someone has a genetic defect does not necessarily mean they will get that particular disease. For instance, inheritance of a mutation in either the BRCA1 or BRCA2 gene accounts for approximately 5 – 10 percent of all breast cancer cases, and varies by family and culture[9-11]. This means 90 - 95% of the carriers of this gene *do not* get the disease, suggesting that a person's internal or external physiological environment, as well as their culture, can play a role in how the breast cancer gene expresses. If only 5 to 10 percent of breast cancers are the result of BRCA1 and BRCA2 mutations in these genes, a particular mutation influences the risk for getting breast cancer depends on other risk factors you may already have[12]. For example, if 10 or more people in various generations of your family have had breast cancer, a particularly dangerous BRCA1 mutation could give you as much as an 85 percent chance of developing the disease by the age of 70. But if you have had only a few relatives with breast cancer, such a mutation probably gives you at most a 56 percent chance of a breast cancer diagnosis before you turn 70. Other significant factors in determining how this gene expresses include co-morbidities and the ability to cope with and handle stress. A genetic counselor can help to sort out whether testing should be done, and your doctor can help you determine whether you have enough of a genetic predisposition to warrant taking medication as a preventative. How well you eat, exercise and handle stress all play a nominal role in the development of disease.

The Human Genome Project began in 1987 and was completed in 2001. Upon its completion it was discovered that genes do not necessarily control the characteristics of life, if you remove the genes from a cell, the cell still lives. Genetic changes often come from external stimuli. Protein receptors on cell membranes have antennae like structures that read signals from the environment and signal the cell of its environmental status. Effector proteins then send signals into the cells to activate the proteins to move towards or away from the signal depending on whether its environment is safe or harmful. This phenomenon, known as epigenetics, can externally influence genetic expression. Epigenetics was coined by C. H. Waddington in 1942 to describe the differentiation of cells from their initial totipotent state in embryonic development. Totipotency is the ability of an embryonic stem cell to differentiate into any other cell in the body. When Waddington coined the term epigenetics, the physical nature of genes and their role in heredity was not known. Waddington used it as a conceptual model of how genes might interact with their surroundings to produce a phenotype. The molecular basis of epigenetics involves modifications of the activation of certain genes, but not the basic structure of DNA. Chromatin proteins associated with DNA can be activated or silenced depending on stimuli in the cell's external environment. This accounts for why the differentiated cells in a multi-cellular organism express only the genes that are necessary for their own activity. Epigenetic changes are preserved when cells divide. These changes have been observed in mice given dietary supplements that have epigenetic changes affecting expression of the agouti gene, which affects their fur color, weight, and propensity to develop cancer[13, 14].

If our genes are not manifesting our disease, what is?

As stem cells differentiate there is potential for cell signaling error when our body regenerates tissue and organs. In diseases such as cancer, one theory is the concept of tumor hierarchy. This concept suggests that a tumor is a heterogeneous population of mutant cells, all of which share some mutations but will vary in specific phenotype. In this model, the tumor is made up of several types of stem cells, one optimal to the specific environment and several less successful lines. These secondary lines can become more successful in some environments, allowing the tumor to adapt to its environment. If the environment is immune compromised (such as in an unhealthy person), the cells can easily form cancer stem cells initiating from tumors. Other types of cell signaling errors can cause viruses to spread, autoimmune diseases to proliferate, and inflammation to spread out of control. Cell communication and cell signaling play a substantial role in how disease manifests.

Disease Manifested Psychologically

One serious mental disorder linked to genetics is schizophrenia. Schizophrenics have an altered sense of reality. People with the disorder may have hallucinations, such as hearing voices or seeing things that are not really there; or delusions, such as believing the authorities will arrest them even though they have broken no laws. This disorder causes them to behave in abnormal ways. Many studies show that there is a strong link between heredity and schizophrenia[15-18]. If an individual identical twin has the disorder, there is a 46 percent chance that the other twin will show symptoms of schizophrenia as well. Children who have two parents with schizophrenia have a 46 percent chance of having the disorder. People who come from families where no one has schizophrenia have only a very small chance (1 percent) of developing the disorder. Genetics is not the only factor that determines whether a person will get schizophrenia. A child whose mother does not have schizophrenia but who is adopted into a family in which one of the adopted parents has the disorder has an 11 percent chance of showing schizophrenic symptoms. Epigenetics and our environment can play a role in the development of schizophrenia[19].

Alcoholism

For many years scientists have known that people with a family history of alcoholism are more likely to become alcoholics themselves. Using twin and adoption studies, researchers have found that genes probably play a part in determining who will become an alcoholic, but as with many behaviors linked to genes, environmental factors are also very important. Researchers are trying to pinpoint genes related to alcoholism, but this is difficult, since it is thought that alcoholism may be caused by many different genes[20] in combination with environmental factors.

Obesity

Obesity is the condition of having much more body fat than is appropriate for a person's age, sex, and height. The chance of becoming obese has been linked to a person's genes as well as environmental factors[21]. Using the methods of inheritance studies, such as investigating pairs of twins, researchers have concluded that inherited genes contribute about 40 percent and the environment about 60 percent to whether a person becomes obese[22]. Although genes are involved in this disorder, environmental factors, such as income level, social eating habits, cultural values, and exercise play important roles.

Depression

Depression is present when a person has feelings of sadness, despair, and hopelessness over a long period of time. Depression is another disorder that has been linked to the genes that a person inherits, but it is a mental disorder that is also closely linked to environmental factors, more so than schizophrenia. It is thought that certain events in life may bring on depression if a person has inherited genes that make it more likely the condition will develop. People who do not have a family history of depression can become depressed. Some of the common environmental factors leading to depression include abuse or neglect, poverty, a traumatic or extremely violent episode in a person's life, and death of a parent or loved one[23].

Can Our Environment Cause Disease?

Epigenetics is the study of heritable changes in gene expression or cellular phenotype caused by mechanisms other than changes in our DNA. Epigenetic variation, whether innate or induced (nature vs. nurture), contributes to variation in gene expression, the range of potential individual responses to internal and external cues, and risk for metabolic disease. Studies using rodent models suggest that during both early development and in adult life, environmental signals can activate intracellular pathways that directly remodel the "epigenome," leading to changes in gene expression and neural function[24, 25]. An example of this is when rat mothers give birth to their pups they will clean and groom them (licking and grooming behavior aka LG), however not all mothers groom their young the same way, causing two groups to form: High LG mothers and Low LG mothers. The high LG mothers groom and lick their young at a much greater rate than the low LG mothers, and as a result it was found that the pups of high LG mothers were healthier, grew at a good rate, and were all around much better off than the pups raised by low LG mothers[26]. These studies define a biological basis for the interplay between environmental signals and the genome in the regulation of individual differences in behavior, cognitive function, and physiology.

In his book *The Biology of Belief,* author Bruce Lipton, a former medical school professor and research scientist, examines in great detail the process by which cells receive environmental information. His research suggests that genes and DNA rarely ever control our biology; but instead DNA is controlled by signals from outside our cells[27]. These include information coming from our thoughts - both positive and negative. Lipton's research is indicative of Candace Pert's findings in her study of neuropeptides[28]. As a neuroscientist and pharmacologist, Pert has published over 250 scientific articles on peptides and their receptors and the role of these neuropeptides in the immune system. She discovered that cells and molecules react much stronger to environmental cues than to genetic instruction.

In order to prevent or offset disease it is important to maintain factors that control the regulation of health and wellness. These factors include our exposure to unhealthy energy, whether physical, mental or emotional. The beginning stage of disease starts with imbalances in our energy field that present identifiable markers that can be manipulated and rebalanced[29-31]. There are three specific events that can cause energy field distortions: 1) physical or emotional trauma,[32]; 2) toxins, such as environmental toxins which interfere with proteins in our body[33]; and 3) cognitive signals from our brain (thoughts) which can lead to disease in our cells[34]. Rebalancing this distortion of energy in our biofield can be done with meditation, energy therapies, appropriate nutrition and physical exercise. Energy therapies can detect imbalances in a person's biofield long before they manifest as disease.

Our biofield permeates our physical body and emanates about 2 – 3 feet beyond the body (Figure 1).

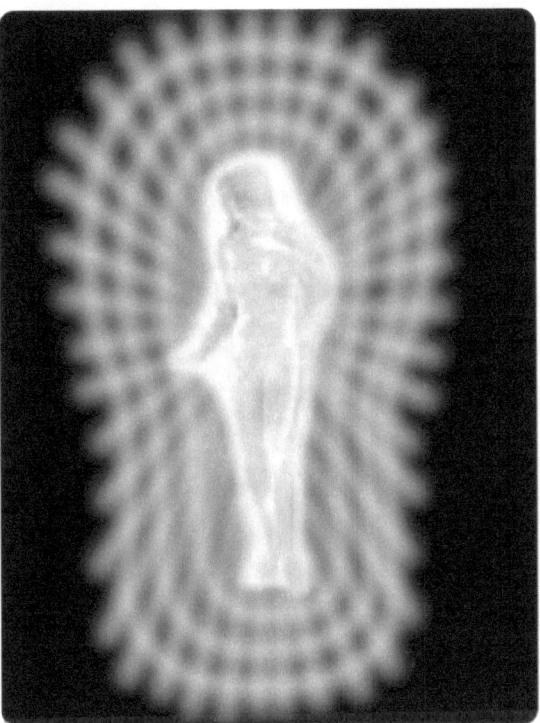

Figure 1: Fields shown in and around the body
appear as 3 dimensional concentric circles

Energy medicine practitioners read these fields by scanning the body with their hands. As a Polarity Practitioner, I have been trained to sense the flow of energy in the human biofield and determine if it is stagnate or blocked. Our body's biofield has measurable electromagnetic emissions which can be detected much the same way electric fields coming from our heart can be measured. Electrical activity of the heart is measured with an electrocardiogram (ECG), and brain waves are measured with an electroencephalogram (EEG). EEG measures voltage fluctuations resulting from ionic current flows within the neurons of the brain[35]. The device used to measure the energy emanating from both the human body and the practitioner's hands is called a Superconducting Quantum Interference Device (SQUID). It is a very sensitive magnetometer used to measure extremely weak magnetic fields, based on superconducting loops containing Josephson junctions. SQUIDs are sensitive enough to measure fields as low as 5 Tesla. A Tesla is a measurement of magnetic flux density. (See figure 2).

Figure 2. Flux is the field lines you see coming from schematics of these magnetic fields.

Practitioners use their hands to feel if the lines are stagnate, blocked or free flowing. Free flowing energy is indicative of health, stagnate energy is indicative of acute illness and blocked energy is indicative of chronic illness depending on how long it has been blocked. People begin to get tired and eventually sick when free flowing energy slows down. Depending on where in the body the energy is blocked, there is a correlative affected with our emotions and psychological health. In a hostile living environment people suffer from more colds, migraine headaches, injure themselves more often, and even become depressed or suicidal. Weak energy in and around the stomach can be indicative of nausea. Nausea is correlated with feeling unsafe. Small concentrated packets of weak energy around the stomach are indicative of ulcers. Dark stagnant energy permeating in and around organs shows the potential for cancer. Cancer is highly correlated with unresolved anger manifested from fear and improperly addressed emotion. Kidney stones are a sign of unresolved anger that is perpetuated through grudges. Ulcers are symptomatic of undue stress or the inability to handle stress. Our reaction to anxiety, strain, worry and tension plays an undeniable role in how our bodies react to stress. When I put my hands in the energy field of people experiencing these emotional issues, the energy blockage is obvious.

We get sick because our energy field is out of balance with respect to nature. Unbalanced energy expresses as stagnation, blockages and incoherent field lines, all of which deplete our biofield of its vital life force. Sickness evolves as feelings of malaise due to a loss of meaning in our life. Disease manifests because of a strong separation between our spirit and our ego. We need our ego to make sure our body receives food, clothing and shelter. Without it our body could not survive; however, as is often the case our ego can run rampant causing us to think that life is only about consumption. There is nothing wrong with an ego that is kept in check, but there must be a balance of spirituality and conscious awareness regarding our existence. When responded to in a productive manner, disease and illness keeps our ego in check. We get sick to create life changes. Superficial living can contribute to a numbing of emotions, remove meaning in life, and cause us to lose our reason to live. Our soul picks up on these cues coming from our mental and emotional body, and the process of dying begins. Death is the release of our soul from our body. We exist for a reason. If there is no purpose for living, the soul finds a way to escape the body. Disease is an exit strategy.

References

1. Beaugerie, L. and J. Petit, *Microbial-gut interactions in health and disease. Antibiotic-associated diarrhoea.* Best Pract Res Clin Gastroenterol, 2004. **18**(2): p. 337–52.
2. Soulsby, E., *Resistance to antimicrobials in humans and animals: Overusing antibiotics is not the only cause and reducing use is not the only solution.* BMJ 2005. **331**(7527): p. 1219–20.
3. Janeway, C., et al., *Immunobiology.* Garland Science Publishing, 2005(6th Edition): p. 19.
4. Temin, H., *Homology between RNA From Rous Sarcoma Virus and DNA from Rous Sarcoma Virus-infected cells.* Proceedings from the National Academy of Sciences, 1964. **52**: p. 323-329.
5. Nijhout, H., *Metaphors and the role of genes in development.* Bioessays, 1990. **12**(9): p. 441-446.
6. Health, N.I.o., *Genetic Testing.* http://www.nlm.nih.gov/medlineplus/genetictesting. html, 2012 Retrieved, April 29.
7. Bellenir, K., *Genetic Disorders Sourcebook.* Omnigraphics, 1996. **Detroit, MI**.
8. http://www.ncbi.nlm.nih.gov/disease, *Genes and Disease—Information and Chromosome Maps from National Institutes of Health.* 2010.
9. Weitzel, J., Lagos, V, Cullinane, C, Gambol, P, Culver, J, Blazer, K, Palomares, M, Lowstuter, K, MacDonald, D, *Limited family structure and BRCA gene mutation status in single cases of breast cancer.* JAMA, 2007. **297**(23): p. 2587-95.
10. Ewald, I., Izetti, P, Vargas, F, Moreira, M, Moreira, A, Moreira-Filho, C, Cunha, D, Hamaguchi, S, Camey, S, Schmidt, A, Caleffi, M, Koehler-Santos, P, Giugliani, R, Ashton-Prolla, P, *Prevalence of the BRCA1 founder mutation c.5266dupin Brazilian individuals at-risk for the hereditary breast and ovarian cancer syndrome.* Hered Cancer Clin Pract, 2011. **20**(9): p. 12.
11. Allain, D., *Genetic counseling and testing for common hereditary breast cancer syndromes: a paper from the 2007 William Beaumont hospital symposium on molecular pathology.* J Mol Diagn, 2008. **10**(5): p. 383-95.
12. Barnes, D., Antoniou, A, *Unravelling modifiers of breast and ovarian cancer risk for BRCA1 and BRCA2 mutation carriers: update on genetic modifiers.* J Intern Med, 2012. **Feb 6. doi: 10.1111/j.**(1365-2796.2012.02502.x. [Epub ahead of print]).
13. Waterland, R., Jirtle, R, *Transposable elements: Targets for early nutritional effects on epigenetic gene regulation.* Molecular and Cellular Biology, 2003. **23**(15): p. 5293–5300.
14. Jablonka, E., Gal, R, *Transgenerational Epigenetic Inheritance: Prevalence, Mechanisms, and Implications for the Study of Heredity and Evolution.* The Quarterly Review of Biology, 2009. **84**(2): p. 131–176.
15. Bergen, S., Petryshen, T, *Genome-wide association studies of schizophrenia: does bigger lead to better results?* Curr Opin Psychiatry, 2012. **Jan 24.**
16. Karege, F., Méary, A, Perroud, N, Jamain, S, Leboyer, M, Ballmann, E, Fernandez, R, Malafosse, A, Schürhoff, F, *Genetic overlap between schizophrenia and bipolar disorder: A study with AKT1 gene variants and clinical phenotypes.* Schizophr Res., 2012. **Jan 23.**
17. Zhang, R., Lu, S, Meng, L, Min, Z, Tian, J, Valenzuela, R, Guo, T, Tian, L, Zhao, W,

Ma, J, *Genetic Evidence for the Association between the Early Growth Response 3 (EGR3) Gene and Schizophrenia.* PLoS One., 2012. **7**(1): p. e30237.

18. Maiti, S., Kumar, K, Castellani, C, O'Reilly, R, Singh, S, *Ontogenetic de novo copy number variations (CNVs) as a source of genetic individuality: studies on two families with MZD twins for schizophrenia.* PLoS One, 2011. **6**(3): p. e17125.

19. Rosenthal, R., *Of schizophrenia, pruning, and epigenetics: a hypothesis and suggestion.* Med Hypotheses, 2011. **77**(1): p. 106-108.

20. Latvala, A., Dick, D, Tuulio-Henriksson, A, Suvisaari, J, Viken, R, Rose, R, Kaprio, J, *Genetic correlation and gene-environment interaction between alcohol problems and educational level in young adulthood.* J Stud Alcohol Drugs, 2011 **72**(2): p. 210-220.

21. Rokholm, B., Silventoinen, K, Ängquist, L, Skytthe, A, Kyvik, K, Sørensen, T, *Increased genetic variance of BMI with a higher prevalence of obesity.* PLoS One, 2011. **6**(6): p. e20816.

22. Hasselbalch, A., *Genetics of dietary habits and obesity - a twin study.* Dan Med Bull, 2010. **57**(9): p. B4182.

23. Althoff, R., Hudziak, J, Willemsen, G, Hudziak, V, Bartels, M, Boomsma, D, *Genetic and environmental contributions to self-reported thoughts of self-harm and suicide.* Am J Med Genet B Neuropsychiatr Genet., 2012. **159B**(1): p. 120-127.

24. Zhang, T.-Y., Meaney, M, *Epigenetics and the Environmental Regulation of the Genome and Its Function.* Annu. Rev. Psychol. , 2010. **61**: p. 439–466.

25. Hellstrom, C., Ian, J. Meaney, M, *Epigenetics and the Environmental Regulation of the Brain's Genome and its Function* Current Psychiatry Reviews, 2010. **6**(14): p. 145-158.

26. Zhang, T.-Y., Meaney, M, *Epigenetics and the Environmental Regulation of the Genome and Its Function.* Annual Review of Psychology, 2010 **61**: p. 439–466.

27. Lipton, B., K. Bensch, and M. Karasek, *Microvessel endothelial cell transdifferentiation: phenotypic characterization.* Differentiation, 1991. **46**(2): p. 117-133.

28. Polianova, M., F. Ruscetti, and C. Pert, *Antiviral and immunological benefits in HIV patients receiving intranasal peptide T (DAPTA).* Peptides, 2003. **24**(7): p. 1093–1098.

29. Coakley, A., Barron, A, *Energy therapies in oncology nursing.* Semin Oncol Nurs, 2012 **28**(1): p. 55-63.

30. Hart, L., Freel, M, Haylock, P, Lutgendorf, S, *The use of healing touch in integrative oncology.* Clin J Oncol Nurs, 2011. **15**(5): p. 519-525.

31. Hardwick, M., Pulido, P, Adelson, W, *Nursing intervention using healing touch in bilateral total knee arthroplasty.* Orthop Nurs, 2012. **31**(1): p. 5-11.

32. Fang, C., Tiao, G, James, H, Ogle, C, Fischer, J, Hasselgren, P, *Burn injury stimulates multiple proteolytic pathways in skeletal muscle, including the ubiquitin-energy-dependent pathway.* J Am Coll Surg, 1995. **180**(2): p. 161-170.

33. von Kleist, L., Haucke, V, *At the Crossroads of Chemistry and Cell Biology: Inhibiting Membrane Traffic by Small Molecules.* Traffic, 2012. Apr 13(4):495-504.

34. Segerstrom, S., *Optimism and immunity: Do positive thoughts always lead to positive effects?* Brain Behav Immun, 2005. **19**(3): p. 195–200.

35. Niedermeyer, E. and F. da Silva, *Electroencephalography: Basic Principles, Clinical Applications, and Related Fields* Lippincot Williams & Wilkins, 2004 **Baltimore, MD.**

CHAPTER THREE
Is There a Way to Prevent Illness?

Since disease is a language, the best way to prevent it or keep it from getting worse is to understand how it manifests. Disease begins at the spiritual level first, emotional level second, mental and cognitive level third and ultimately on the physical level. Because conventional medicine addresses mostly physiological and psychological disease, we will begin with these. Currently disease prevention is described in three phases:

1. *Primary* – directly addresses the mediating causes of disease and how it is carried out before its onset, thereby preventing its occurrence (i.e. weight loss, tobacco and alcohol cessation, healthier eating and exercise habits, etc.).
2. *Secondary* – early detection and treatment of disease before full blown illness develops (i.e. mammography, Pap test, colonoscopy, etc.).
3. *Tertiary* – attempts to prevent recurrence or progression of disease that has already occurred. (i.e. eye, foot and kidney abnormalities for diabetics, smoking cessation for COPD patients, etc.).

These descriptions are good, but incomplete. Since most diseases are classified as *etiology unknown,* meaning "of unknown origin", they are categorized by their symptoms. Because there are few known causes of disease, it is important to look beyond the physiology towards a psychological and emotional cause, ultimately addressing the spiritual. Understanding our emotions is essential to our health and well being. In Victorian societies the expression of emotion was thought to be mentally unstable behavior. Perhaps this is similar to other cultures, but this type of thinking is unfortunate because suppressing our emotions can lead to physical illness[1-3]. One very common emotional disorder is depression, which is a leading cause of physical illness [4,5]. Depression can arise from low self-esteem, anger, boredom or lack of passion, but most the time it is due to traumatic life experience caused by childhood events such as loss or extreme trauma. Chronic depression can be caused by emotional, physical or sexual abuse, yelling or threats of abuse, neglect, constant criticism, inappropriate or unclear expectations, maternal separation, family conflict, divorce, addictive behavior, family violence, racism and poverty[6]. Left untreated these emotional issues continue into adulthood. The onset of depression can be very subtle, becoming more profound as we live our lives amidst unresolved conflict and without purpose. The onset of depression is difficult to detect. It can be masked by alcohol consumption, sleeping pills or numbing behavior in front of the television or computer.

It is estimated that over 118 million people worldwide take some form of antidepressant, and two-thirds are women[7]. Depressive disorders affect approximately 18.8 million American adults age 18 and older or about 9.5% of the U.S. population annually. This includes major depressive disorder, dysthymic disorder (depression associated with fatigue, appetite disorders and insomnia), and bipolar disorder[8]. About one in 10 Americans aged 12 and over takes anti-depressant medication[9]. Overall, females are 2½ times as likely to take antidepressant medication as males. However, there is no difference by sex in rates of antidepressant use among persons aged 12–17. Twenty-three percent of women aged 40–59 take antidepressants, more than in any other age-sex group. There is no difference by income in the prevalence of antidepressant usage. Fifty-seven percent of Hispanic and African Americans however report using some form of Complementary and Alternative Medicine to combat depression[10]. There may be a genetic basis to some depression, but studies show it is mainly associated with a stressful or traumatic event, with trauma preventing certain parts of the brain (hippocampus and frontal lobe) from functioning properly[11]. Recent studies indicate that the stress hormone cortisol is more of a factor in depression than serotonin, which is targeted by most antidepressants (Selective Serotonin Reuptake Inhibitors or SSRIs)[12]. Depression has been found to speed up the decline of cancer patients[13]. Twenty-six studies with a total of 9417 patients examined the effects of depression on patients' cancer progression and survival. In these combined studies, the death rates increased 25 percent in patients experiencing depressive symptoms, and death rates were 39 percent higher in patients diagnosed with acute or chronic depression. The risks were measured separately from patients' other clinical symptoms affecting survival, suggesting that depression may actually play a part in shortening survival time.

In his book, *Artificial Happiness: the dark side of the New Happy Class*, author Ronald Dworkin writes about the politics of health care and medical ethics that have changed the way patients and doctors relate to societal norms. Dworkin coins the term "artificial happiness", which he describes as the increase of prescribed psychotropic drugs. He suggests that alternative medicine and the upregulation of endorphins might be a better way to maximize health through exercise and spiritual practice which can help people feel happy despite their depression. Dworkin identifies a dark side of coping, suggesting that Prozac, for instance, is keeping people in unsatisfactory jobs and relationships, numbing their impulse to make changes necessary to overcome depression, because of the artificial happiness induced by such drugs[14].

As much as chronic depression can be very demanding, mild depression can be easily ignored. Instead of addressing the problems causing depression and making the appropriate changes, we reach for medication to ease our discomfort. Ignoring mild depression can lead to chronic depression. Unchecked emotional instability can also cause eating disorders[15, 16], obsessive compulsive disorder[17], anxiety disorder[18], and even bipolar disorder[19]. If acute illness goes untreated, chronic illness appears. For example malnutrition leads to immune deficiency and inability to survive[20], and untreated bacterial infection can lead to sepsis[21]. Detecting and reversing the onset of disease begins by responding to our emotions and relating uncomfortable feelings to physical symptoms. Noticing when we feel sad, angry, irritated, frustrated and even bored is as important to detecting disease as are signs of runny nose and headache before the onset of a cold.

Indications of the onset of more serious diseases include severe sinus infections, flu, or irritable bowels. Bouts of acid reflux, constipation, diarrhea and headaches are also indicative of more serious conditions yet to come. While using over the counter remedies temporarily alters the problems, many of these acute problems can be permanently reversed by changing how we react to our environment, especially stress. Stress is a major cause of acute as well as chronic illness[22, 23]. Mental stress has been shown to increase the risk of heart disease. Researchers studied the perceived stress levels of more than 73,000 Japanese men and women ages 40-79 whose health was monitored for eight years. Nearly 9,000 women and 7,000 men reported high mental stress.

Researchers found that women in the high-stress group were more than twice as likely as those with low stress to suffer a stroke or develop heart disease[24]. This study accounted for other potential causes. High-stress women were more likely to show a history of high blood pressure or diabetes than women with less perceived stress. The high-stress women tended to be younger, more educated, and less physically active, and also more likely to be angry, in a hurry, feel hopeless, and unfulfilled. Men who reported medium or high levels of mental stress were nearly twice as likely to suffer a heart attack. How people perceived and coped with stress was the biggest factor in whether or not their mental state led to physical symptoms.

We All have Stress – how do we cope with it?

Besides pharmaceuticals there are many ways to change the way we react to our environment to help us cope with stress. We innately sense whether we are safe or in danger. Input from our environment is perceived through our nervous system, which is a network of nerves that includes the brain and spinal cord. Our nervous system responds to external as well as internal stimuli. Our nerves (neurons) use electrochemical signals to give specific instructions to various parts of our body as to how we should react to specific stimuli. Stress stimulates our central nervous system and prepares it to meet stressful situations. During the preparation, our body goes through various physiological changes that enable us to either fight the enemy or try to escape the environment we perceive as threatening. These physiological changes are the function of the autonomic nervous system, which is made up of the sympathetic and the parasympathetic nervous systems. These are opposing systems. The sympathetic nervous system is responsible for fight or flight, while the parasympathetic nervous system prepares the body for rest and relaxation. Our central nervous system, as well as our adrenal glands, can be pushed into overload by challenging stressful environments. Mental and emotional stress can put us into a constant fight or flight mode that can be difficult to switch off. How we react to events in our environment, whether stressfully or calmly, affects us physically. If our perception is internalized as stressful, we may always feel stressed out in that same situation despite any external threat - whether real or imagined. This internalization process can damage both our nervous and hormonal systems[25]. When we perceive a threat, our nervous system responds by releasing a flood of stress hormones, including adrenaline and cortisol. These hormones alert the body to react to a threat. The stress response is the body's way of protecting us. When working properly it helps us stay focused, energetic, and alert, but chronic stress can significantly damage our immune system[26, 27].

The term psychoneuroimmunology describes the interaction between our mental state, nervous and immune systems, and the interconnections of these systems. Immune system changes create more susceptibility to infection, and have been known to increase the potential for an outbreak of psoriasis for people with that skin disorder[28]. A chronic over-stimulation of our sympathetic nervous system can lead to suppressed immunity and adrenal exhaustion[29]. When an event causes fear, trauma, or dreaded situation, the sympathetic nervous system senses danger and increases the heart rate sending an extra supply of blood to different parts of our body. It also signals the adrenal glands near our kidneys to secrete adrenaline, the hormone that sends energy to our muscles. After the increase or upregulation of these hormones prepares our bodies for fight or flight, our nervous system will then down regulate or reduce the level of hormones in the blood stream, signaling our heart rate to return to normal. This is a good system if we are about to be in a car accident, to get burned in the oven, or get hit with an object. But if our sympathetic nervous system is constantly in the "on" position, it can be detrimental to our health. The parasympathetic nervous system cannot function at the same time as the sympathetic nervous system, therefore relaxation is never initiated. Emotional stress produces similar physical changes by stimulating the nervous system to respond. Constant stress puts the sympathetic nervous system in a constant state of alert, not allowing the body to ever relax and rejuvenate.

While it is impossible to totally avoid stress, our reaction to it and the way we handle it is the embodiment of disease prevention. Deep breathing, yoga and meditation send signals to the central nervous system that initiate our return to parasympathetic or relaxed state[30, 31]. Changing our perception or changing how we react to certain situations also affects us physiologically. Change occurs by visualizing the outcome we would like to experience. The mind has a difficult time differentiating between perception and reality[32]. To create a relaxed state in our body simply focus on a calm environment in our mind. Albeit difficult at times, meditation is vital to our health and well being. Noticing our stress levels and appropriately changing how we react to our environment is extremely important to our physical health. Our very survival depends on our being able to connect and interact with our environment in a safe and healthy manner. If we are consciously taking inventory of our health for the sake of avoiding or reversing disease, then perhaps it would be worth learning to quiet our minds and control our reaction to stressful events. Decreasing stress can reverse acute phase disease states and improve the quality of our life even if we are experiencing chronic or incurable disease[33].

Healthy eating habits and lifestyle are excellent steps to disease prevention. There is energy in food, there is energy in exercise. Food is the energy that fuels our body and affects our mind and even our soul. The effect of this energy on our body is the basis of macrobiotics. Macrobiotics is based on the view that we are the result of and are continually influenced by our total environment, including the foods we eat, our daily social interactions, the climate and geography in which we live and how much we move and exercise our body. Macrobiotics views sickness as the natural attempt of the body to return to a more harmonious and dynamic state with the natural environment. Healthy energy exists on a continuum - from *yin,* among whose attributes are cold, darkness, and contraction, to *yang,* characterized by heat, light, and expansion. Foods exist on this continuum and are characterized as predominantly yin or yang according to the effect they have on the body.

Yin foods — primarily sweeteners, oil, liquids, and most dairy products which have a cooling and stagnating affect on us. Overconsumption of these foods disperses internal energy and is a primary cause of the weakness and degenerative diseases[34]. Caffeine, alcohol, and many drugs are also in this category. Although they along with artificial sweeteners produce an initial burst of energy, their effect quickly slows down and is followed by a depletion of energy. Because their overall effect is slowing and cooling, they are classified as yin.

Yang foods — primarily meat, salt, eggs, and hard cheeses — have a heating and animating effect on the body. Overconsumption results in tension and rigidity. Regular consumption of either yin or yang foods leads to imbalances of our body's energy that ultimately manifests as physical problems. There are yin diseases and yang diseases, characterized by an excess of one or the other, and diseases resulting from both. It is best to eat foods more toward the center of the food-energy spectrum. These include fresh, organic fruits and vegetable and lean meats.

Exercise produces energy in our body. Our exercise regime should complement our lifestyle, instead of mirror it. For instance, if we run all day in a high pace, high energy job, our exercise should be more paced and fluid, such as swimming, bike riding and yoga. In contrast if we sit all day, our exercise routine should involve more rigorous routines such as jogging, tennis, basketball, or dance. All people need resistance training for muscle building and stamina. Without weight lifting or resistance training muscles can atrophy and bones become brittle. A review of 12 large-scale studies on the connection between exercise and fatigue was published. The studies took place from 1945 to 2005, and each study measured the amount of physical activity that participants were doing and how much energy or fatigue the participants experienced. All of the studies found a direct link between reduced risk of fatigue for people who were physically active compared to those who were inactive. Other research shows that even among people with chronic illness like cancer[35], heart disease[36], and cognitive impairment[37], exercise can ward off feelings of fatigue and help people feel more energized. Meditation, healthy foods and exercise are all ways in which we can reduce stress to prevent illness. Our survival is predicated on the ability to accurately assess and respond to environmental information[38]. Learning to cope with stress not only prevents the onset of disease but can improve our quality of life (QOL) if currently experiencing disease.

References

1. Harris, W., et al., *Circulatory and humoral responses to fear and anger.* The Physiologist, 1964. **7**: p. 155.
2. DiGiuseppe, R. and R. Tafrate, *Understanding Anger Disorders.* Oxford University Press:London, 2006: p. 133-159.
3. Hochschild, A., *The managed heart: commercialization of human feeling.* University California Press 1983.
4. Trivedi, M., *The link between Depression and Physical Symptoms.* Prim Care Companion J. Clin Psychiatry, 2004. **6**(suppl 1): p. 12-16.
5. Nunez, F., A. Vranceanu, and D. Ring, *Determinants of pain in patients with carpal tunnel syndrome.* Clin Orthop Relat Res, 2010. **468**(12): p. 3328-3332.
6. Glaser, D., *Child Abuse and Neglect and the Brain.* J Child Psychol. & Psychiat, 2000. **41**(1): p. 97-116.
7. http://www.cdc.gov/features/datastatistics.html#Dec2007, *Antidepressants.* 2010.
8. NIMHstatistics@mail.nih.gov., *National Institutes of Mental Health.* 2010.
9. Prevention, C.f.D.C.a., *National Center for Health Statistics Data Brief.* http://www.cdc.gov/nchs/data/databriefs/db76.pdf, 2011. **76**.
10. Bazargan, M., et al., *Correlates of complementary and alternative medicine utilization in depressed, underserved african american and Hispanic patients in primary care settings.* J Altern Complement Med., 2008. **14**(5): p. 537-44.
11. Vythilingam, M., *Childhood Trauma Associated With Smaller Hippocampal Volume in Women With Major Depression.* American Journal of Psychiatry, 2002. **159**: p. 2072-2080.
12. Burns, D. and W. Danton, *Rumble in Reno: The Psychosocial Perspective on Depression.* Psychiatric Times, 2000. **17**(8): p. 138.

13. Satin, J., W. Linden, and M. Phillips, *Depression as a predictor of disease progression and mortality in cancer patients: a meta-analysis.* Cancer, 2009. **115**(22): p. 5349-5361.

14. Dworkin, R., *Artificial Happiness: the dark side of the new happy class.* Carroll & Graf Publishers, 2006. **New York: NY.**

15. http://www.win.niddk.nih.gov/publications/binge.htm, *Binge eating disorder.* Weight-control Information Network, 2008.

16. http://tinyurl.com/25f6v76, *Treatment of eating disorders.* National Eating Disorders Association, 2010.

17. http://www.nimh.nih.gov/health/topics/obsessive-compulsive-disorder-ocd/index. shtml, *Obsessive-compulsive disorder, OCD.* NIMH, 2010.

18. www.nimh.nih.gov/health/publications/anxiety-disorders/complete-index.shtml, *Anxiety disorders.* NIMH, 2010.

19. http://www.nimh.nih.gov/health/topics/bipolar-disorder/index.shtml, *Bipolar disorder.* NIMH., 2010.

20. Felblinger, D., *Malnutrition, infection, and sepsis in acute and chronic illness.* Crit Care Nurs Clin North Am., 2003. **15** (1): p. 71-78.

21. Becker, J., et al., *Surviving sepsis in low-income and middle-income countries: new directions for care and research.* Lancet 2009. **Infect Dis.**(9): p. 577-582

22. Korenromp, I., Grutters, J, van den Bosch, J, Heijnen, C, *Post-inflammatory fatigue in sarcoidosis: Personality profiles, psychological symptoms and stress hormones.* J Psychosom Res, 2012. **72**(2): p. 97-102.

23. Wong, D., Tai, T, Wong-Faull, D, Claycomb, R, Meloni, E, Myers, K, Carlezon, W, Kvetnansky, R, *Epinephrine: A Short- and Long-Term Regulator of Stress and Development of Illness : A Potential New Role for Epinephrine in Stress.* Cell Mol Neurobiol, 2011. **Nov 17.**

24. Iso, H., et al., *Percieved Mental Stress and mortality from cardiovascular diseae among Japanese men and women.* Circulation: Journal of the American Heart Assocation, 2002. **106**(1229-1236).

25. Tsigos, C. and G. Chrousos, *Hypothalamic-pituitary-adrenal axis, neuroendocrine factors, and stress.* Journal of Psychosomatic Research, 2002. **53**: p. 865–871.

26. Jordan, N., et al., *Systemic mycobaterium avium complex infection during anti- TNF-alpha therapy in a pediatric patient with crohn's disease.* J Pediatr Gastroenterol Nutr., 2011(June 17).

27. Reiche, E., Nunes, S, Morimoto, H, *Stress, depression, the immune system, and cancer.* Lancet Oncol, 2004. **5**(10): p. 617-25.

28. Ron de Kloet, E., M. Joels, and F. Holsboer, *Stress and the brain: from adaptation to disease.* Nature Reviews Neuroscience, 2005. **6**(6): p. 463–475.

29. McCance, K., *Pathophysiology, The Biological Basis for Disease in Adults and Children.* 3rd Ed, 1998. **Mosby Elsevier Health Science.**

30. Jellesma, F. and J. Cornelis, *Mind Magic: A Pilot Study of Preventive Mind-Body-Based Stress Reduction in Behaviorally Inhibited and Activated Children.* J Holist Nurs., 2011. **25**: p. 252.

31. Veerabhadrappa, S., et al., *Effect of yogic bellows on cardiovascular autonomic reactivity.* J Cardiovasc Dis Res, 2011. **2**(4): p. 223-227.

32. Abraham, A. and D. von Cramon, *Reality = Relevance? Insights from Spontaneous*

Modulations of the Brain's Default Network when Telling Apart Reality from Fiction. PLoS ONE, 2009. **4**(3): p. e4741.

33. Smith, M., Goodfellow, L, *The relationship between quality of life and coping strategies of adults with celiac disease adhering to a gluten-free diet.* Gastroenterol Nurs, 2011. **34**(6): p. 460-468.

34. Lerman, R., *The macrobiotic diet in chronic disease.* Nutr Clin Pract., 2010. **25**(6): p. 621-626.

35. Basen, K, Carmack, C, Blalock, J, Baum, G, Rahming, W, Demark, W, *Predictors of Cancer Survivors' Receptivity to Lifestyle Behavior Change Interventions.* Cancer Epidemiol Biomarkers Prev, 2012. **Feb 15. [Epub ahead of print]**.

36. Ekelund, U., Luan, J, Sherar, L, Esliger, D, Griew, P, Cooper, A; International Children's Accelerometry Database (ICAD) Collaborators, *Moderate to vigorous physical activity and sedentary time and cardiometabolic risk factors in children and adolescents.* JAMA., 2012. **15**(307): p. 704-712.

37. Denkinger, M., Nikolaus, T, Denkinger, C, Lukas, A, *Physical activity for the prevention of cognitive decline : Current evidence from observational and controlled studies.* Z Gerontol Geriatr, 2012. **45**(1): p. 11-16.

38. Lipton, B., Bhaerman, S, *Spontaneous Evolution: our positive future (and a way to get there from here).* Hay House: Carlsbad, CA, 2009: p. 255.

CHAPTER FOUR
How Can We See Disease Coming?

It is difficult to see disease coming because we are unfamiliar with how our soul manifests disease physically. We are born totally focused on our spiritual self, but as we learn to become more human we separate the high frequency vibrations of our spirit and focus more on lower vibrations of our physical being. Sometimes we are able reconnect with spirit which is referred to as a "mountain-top experience", where for a brief moment we are so connected to the earth, to God, to all of humanity, we are consciously aware of our soul. Unfortunately these moments are rare, so in order to understand who we are spiritually, it is easier to study the construct of disease from an emotional standpoint. It is unfortunate that we often quiet our emotional pain and anxiety with pharmaceuticals, ignoring the information it is giving us. We compartmentalize who we are, going to different specialists depending on our problem, leaving our spirit to communicate through our physiology instead of through our conscious mind. By the time disease has manifest, its message has been so far removed from our awareness that we cannot see how acute health problems such as rashes, pimples, headaches and eye problems are interrelated to chronic health problems such as stomach and liver disorders. These diseases are all indicative of energetic instability in our biofield, and are the result of unresolved anger, disappointment, and lack of self-esteem. Conditions such as edema, hip dysplasia, breast cancer and foot disorders are the result of unresolved deep seated emotional trauma (both physical and mental), as well as relationship issues. Bowel and bone disorders have a common connection that are the result of not feeling safe, having inappropriate boundaries, or having family problems.

We do not see disease coming because we do not understand disease in a holistic, integrative way. Every cell in our body is constantly moving toward equilibrium as an internal regulatory mechanism. Because the internal and external environments of our body are constantly changing, continuous adjustments must be made to equalize our biofield. Energetic homeostasis is an attempt to maintain our internal condition by limiting fluctuations. It involves a series of negative feedback loops, therefore if one aspect of our energy becomes contracted another will expand and vice versa.

What Signs Should We Look for as Evidence of Disease?

Energy practitioners have discovered links between the lack of energy flow in the body and physical disease. Emotions affect our body's immune system. Table 1 shows the correlation of emotional imbalance to physical disease.

Emotion	Mental State	Acute Illness	Chronic Illness	Incurable Disease
Fear	Avoidance Lack of courage Anxiety Stubborn behavior Stuck in old patterns	Panic attacks Elevated blood pressure Constipation Diarrhea	Allergies/ Eczema Irritable Bowel Syndrome Chronic muscle tension	Crohn's Disease Colon cancer Anal cancer Rectal cancer
Stress	Perception of threat Jealousy Lack of compassion Stagnated thoughts	Fainting spells Elevated respiration, triglycerides and blood pressure	Ulcers Chronic pain Fibroids Chronic anxiety Heart attacks	Ovarian and cervical cancers Autoimmune disease Prostate cancer Lung cancer
Grief	Crying Lack of focus Close-mindedness	Insomnia Restlessness Heartburn	Osteoarthritis Thyroid diseases	Cardiac disease Diabetes Stroke
Anger	Rage problems Blaming others Egotism	Inflammation Acne, boils, and infection, fever	Migraine headaches Acid reflux Cysts and tumors Glaucoma	Myocardial infarction Stomach and liver disease Blindness
Sadness	Indecisiveness Compulsiveness	Overeating Indigestion	Edema Obesity	Kidney disease Breast cancer

Emotion	Mental State	Acute Illness	Chronic Illness	Incurable Disease
Depression	Negative self-image Suicidal thoughts Feeling spaced out Brain fog	Fatigue Joint aches Sleep deprivation Back pain	Fibromyalgia Circulatory problems Arthritis	Bell's palsy Alzheimer's disease Dementia

Table 1. Evolution of disease from the mental and emotional
bodies to the acute, chronic and incurable disease states

Emotions are neither good nor bad they are simply information informing us to whether our biofield is stable. Fear, anger, boredom, grief, depression, and loss of passionate productivity are signs of the onset of disease. Emotional imbalances such as constant crying or wanting to cry, lack of motivation to go to work or perform every day activities are also signs of the onset of disease. Acute illness such as constant colds, cold sores, flu, aches and pains are indicative of illness shifting from psychological/emotional state to physical illness – the tertiary phase of disease. Instead of taking painkillers to mute the information informing us something is wrong, it would be healthier to embrace the information and alter our environment or our reaction to it. Reducing stress goes a long way towards changing our emotional response[1]. Stress reduction can be attained through meditation, slowing the breath, yoga, drinking pure spring water, and eating a healthier diet. These practices have higher vibratory frequencies which increase the flow of energy throughout our body - promoting health and wellness instead of sickness and disease.

How Does Disease Communicate?

Disease communicates through vibration. It uses different frequencies to relay messages that either make us feel pleasant or uncomfortable. These signals can be either harmonic and peaceful or chaotic and uncomfortable. Valerie Hunt, PhD, Professor Emeritus in the UCLA Department of Physiological Sciences has been conducting research in this field for over 40 years. She was the first to research the relationship between changes in bioenergy fields and human behavior. In mapping bioenergy fields, Hunt has found that each individual has a unique resting pattern she calls the Signature Field. "The Signature Field of a healthy human being is composed of balanced, coherent energy patterns running the full spectrum of frequencies (4 – 20 microns in wavelength[2]). This coherency shows up on a graph as smooth, gentle, shallow waves evenly distributed throughout the frequency spectrum" (Figure 1).

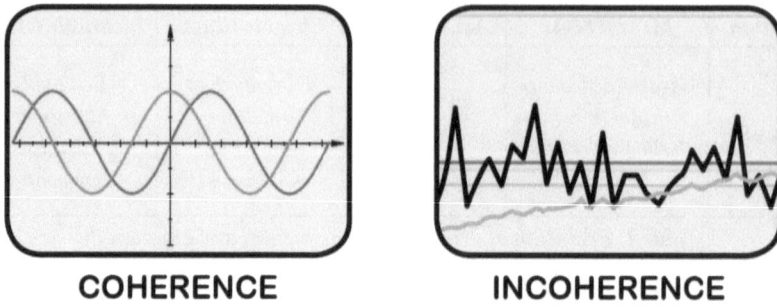

COHERENCE   INCOHERENCE

Figure 1. Coherence is smooth and harmonic while incoherence is static noise

There are two types of patterns in the Signature Fields of people who have (or are soon going to develop) disease: deficiency patterns, and hyperactive patterns. They appear in graphs as thick, jagged waves concentrated in the high- or low-frequency bands. Deficiency diseases like cancer and chronic fatigue syndrome have what Hunt calls "anti-coherent" patterns in the high frequency ranges, with almost no energy at all in the lower frequencies. Hyperactive conditions like colitis, hypertension, and skin problems show anti-coherent patterns in the low frequencies, with absent vibrations in the high frequencies[3]. Through the changes in healthy patterns, Hunt has been able to determine the onset of disease before it manifests in the body.

Everything vibrates. Every atom in our body has a specific vibratory or periodic motion. Each periodic motion has a frequency (the number of oscillations per second) measured in Hertz (Hz). Every element in the chemical Periodic Table has a specific vibratory frequency generating waves that travel out from their source, transferring energy to objects they pass through. Molecules have vibratory frequencies, cells have vibratory frequencies, tissues have vibratory frequencies, organs have vibratory frequencies, *human bodies have vibratory frequencies*...these frequencies are the basis for the human biofield.

Bruce Tainio and Gary Young performed research studies indicating the normal frequency range of the human body is between 62-68 MHz; but if it drops below 62 MHz, the individual becomes a candidate for illness. Cold symptoms appear at 58 MHz, flu symptoms at 57 MHz, Candida at 55 MHz, Epstein bar at 52 MHz, cancer at 42 MHz[4]. In order to nourish and regenerate cells, organisms use enzymes to break down molecules. These enzymes have specific vibratory frequencies[5, 6]. Our cognitive thoughts and our conscious minds also have vibrational frequencies[7]. Tainio and Young showed a negative mental state lowers body frequency by 10-12 MHz.

Subsequently a positive mental attitude, prayer or meditation was shown to raise body frequency by 10-15 MHz. Higher frequency can increase a lower frequency due to the principle of entrainment, which is the tendency for two oscillating bodies to vibrate harmonically. Vibrations oscillating at various frequencies produce noise known as chaotic or incoherent frequencies. These incoherent frequencies can be detected and corrected at the most subtle levels (1-30 Hz). These subtle frequencies are known as sub extremely low-frequency electromagnetic fields (sub ELF-EMF), and are the frequencies at which energy medicine practitioners produce therapeutic results[8].

The information given us by disease can be evaluated at any step in the disease process. Even if an incurable disease has manifested, steps can be taken to improve our quality of life and promote health and wellness. Energy medicine modalities can interrupt a disease process at any stage of its manifestation and can also be used as preventative medicine[9].

The Language of Disease

1. The language of disease starts within our spirit body, years of separation between our spirit and ego create more dependence on understanding our emotions to interpret the information coming from our soul.
2. Be alert for overwhelming anxiety or dissatisfaction with life, these are the primary stage of disease.
3. Pay attention to emotional instability such as boredom, depression, fear, and anger, these are secondary stages of disease.
4. Signs of acute illnesses like headaches, acid reflux, stomach cramps or increased accidents are tertiary stages of disease which move into chronic phase and ultimately to incurable states.
5. Ignoring the language of disease by taking drugs to make the "talk" go away suppresses information. Choose instead to alter the body's environment through meditation, yoga, healthy food choices (food is medicine) and passionate productivity, promoting parasympathetic nervous system response as opposed to sympathetic nervous system response. [always see a physician before discontinuing any medications]
6. Muting the language of disease can cause mild headaches to become chronic (perhaps migraines), acid reflux to become ulcers, and stomach cramps to become irritable bowel syndrome. Instead of

increasing drug dosages or taking additional drugs, try interpreting the message of the disease.

7. Chronic stage disease will ultimately bring on incurable disease states, the ultimate act of communicating the soul's desire to no longer exist in human form.

8. The language of disease is communicated through vibration. Different frequencies express various types of information.

References

1. Robins, C., Keng, S, Ekblad, A, Brantley, J, *Effects of mindfulness-based stress reduction on emotional experience and expression: a randomized controlled trial.* J Clin Psychol, 2012. **68**(1): p. 117-131.

2. Rubik, B., *The Biofield Hypothesis: its biophysical basis and role in medicine.* J Altern Complem Medicine, 2002. **8**(6): p. 703-713.

3. Hunt, V., *Infinite Mind: Science of the Human Vibrations of Consciousness.* Malibu, CA: Malibu Publishing Co, 2000.

4. Young, G., *Frequency of essential oils.* AROMATHERAPY The Essential Beginning, 1996. **1** p. 90, 139.

5. Formaggio, F., Toniolo, C, *Electronic and vibrational signatures of peptide helical structures.* Chirality., 2010. **22**(Suppl 1): p. E30-39.

6. Yu, L., Greco, C, Bruschi, M, Ryde, U, De Gioia, L, Reiher, M, *Targeting intermediates of [FeFe]-hydrogenase by CO and CN vibrational signatures.* Inorg Chem, 2011. **50**(9): p. 3888-3900.

7. Reimers, J., McKemmish, L, McKenzie, R, Mark, A, Hush, N, *Weak, strong, and coherent regimes of Fröhlich condensation and their applications to terahertz medicine and quantum consciousness.* Proc Natl Acad Sci USA, 2009. **106**(11): p. 4219-4224.

8. Oschman, J., *Energy Medicine: The Scientific Basis.* Churchill Livingstone, 2000. **Edinburgh**.

9. Anderson, J., Taylor, A, *Effects of healing touch in clinical practice: a systematic review of randomized clinical trials.* J Holist Nurs., 2011. **29**(3): p. 221-228.

CHAPTER FIVE
Is it my Fault I Got Sick?

Guilt and shame have no place in the healing process. This type of thinking goes back to the idea we deserve to be punished for something. These are insecure and unhealthy thoughts that are of no benefit to us as divine and powerful individuals. Disease is information. When we do not understand a language we do not know how to interpret what is happening. Once we learn the language of disease there is an awakening, not only in our minds, but in our bodies. If you have seen the 1962 movie *The Miracle Worker* about the life of Helen Keller, you will recall the moment when a young, blind, deaf and mute Helen (portrayed by Patty Duke), eventually realizes what her teacher Annie Sullivan (played by Anne Bancroft) is trying to communicate using hand gestures. Once she realizes the potential of this form of communication she runs to every object in her environment asking Annie to sign to her the name of that object. As she begins to understand what language is, her whole world opens up. At last she realizes she has the ability to understand who she is with respect to objects around her. This analogy expresses the excitement our soul feels once it is being understood by our conscious mind. We in turn will want to know everything our soul, our psyche, our emotions and our body is revealing through the language it speaks. Disease is only one language it speaks. Dreams are another language it speaks. Disease is an opportunity to gain information. Fear and dread will make us into victims, while understanding empowers us.

Disease 101

As we discussed in the previous chapter disease starts deep within us – at the soul level. Understanding why we incarnated in human form, our soul's purpose or life's work will fill us with vital energy and a reason for living. Active involvement in our own healing process underscores the need to live a happy and peaceful life. Meditation is the fastest and least expensive way of achieving peace. Many people find it easier to meditate near a water source because it is relaxing. If you do not have ready access to a water source imagine being near one. Whether you are attracted the beach, a lake, a river or waterfall, choose the scene that makes you feel the most peaceful (if you are afraid of water, think of a peaceful place at home). Once you have closed your eyes and visualized this peaceful environment, take a deep breath. Breathing deeply involves your entire lung capacity[1]. Most people use only about one-fourth of their lung capacity. Using your diaphragm, expand your rib cage out as far as possible, hold your breath for 4 counts then exhale slowly. Immediately the muscle tension surrounding our skeleton will release and our heart rate will decrease. Relaxation decreases heart rate and blood pressure[2], and breathing returns to normal. This behavior activates our parasympathetic (PNS) nervous system. When relaxed we think more clearly and digest our food more efficiently as well as sleep more deeply[3]. Once we have achieved a peaceful state we increase our ability to meditate. It is important to meditate on what makes us happy. There are many things that bring people happiness. By far the number one response when asked "What brings you happiness?" is family and friends[4]. Family and friends make it possible for us to express ourselves in ways we could not without being in these relationships. Happiness also comes from finding our passion, our reason for living, our life's work. Finding our passion involves a driving interest in a subject that brings joy with every thought of it. Our passion often involves personal productivity, especially when it aligns with the greater good of the planet. Self absorbed and self-serving thoughts are very different than passion. Life's work can mean sharing our passion for plants, animals, people, the environment, poetry, painting, sculpting, dancing, writing, science, medicine, politics, or religion. On the happiness scale, meaningful work, positive thinking, and the ability to forgive come in close second, third and fourth in how people respond to what makes them happy[4]. What does not seem to make people happy are money, material possessions, intelligence, education, age, gender or attractiveness. I have found in my healing practice that the happiest people are also some of the healthiest. They are people who see a higher purpose in their existence.

Finding purpose in life helps us fend off illness and promote health and well being[5]. Finding passion involves reemerging with our soul and is not a trivial step, but important in overcoming the onset of disease. Understanding how disease begins means understanding how our soul interprets disharmony. The quickest way to know if our soul is becoming distressed is to examine it through our emotions. At the very least we should wake up calm and rested. If this is not happening then it is necessary to determine why.

Besides happiness, peace is essential to preventing or intensifying illness. Stressful reactions to our environment are often prompted by relationships in our lives. Parents, in-laws, children, pets, bosses, co-workers and neighbors can disrupt the peace in our lives. Other issues include financial matters, lack of productivity, and grief caused by the death of a loved one. Some of these situations we have no control over, others we do. For the situations we have no control over, such as who we are related to, there is little we can do; we can however change the way we react to people in these relationships. Most stress comes from our reaction or over-reaction to any given situation. One way to avoid over-reacting is to feel peaceful and happy as often as possible. There are wonderful CDs that play peaceful water sounds and DVDs of peaceful water environments. Ten minutes per day meditating on these peaceful sounds will calm our mind and emotions, initiating the healing process. Spas and wellness centers understand this concept well. They have water features, peaceful music, aromatherapy and white noise that create healing environments. Fortunately we can recreate this same healing spa experience in our own home. Self care is a large part of self-healing. People around us benefit when we are peaceful and happy.

Watching for Signs of Disharmony to Prevent Disease

Creating a peaceful environment is more easily accomplished at home, however when we get to work everything seems to return to the uncontrolled chaos that stressed us in the first place. If you work from home or are a caregiver to children or older or disabled family members, by necessity you must have a space and time to yourself, even if only a few moments a day. Mastering our environmental stress level begins with conscious breathing. When we are stressed we tend to hold our breath or breathe in a shallow manner[6]. Practice breathing meditation on the way to work. If you are calm when you walk into work, but get stressed when you see a co-worker or boss, practice meditative breathing during the meeting or immediately before and afterward. There are several methods to practicing meditative breathing[7]. My favorite is a deep inhale, held for 4 full seconds, then slow exhale. Controlled breathing will help regulate stress levels by controlling our reaction to it[8]. State of mind is key to how our body manifests disease[9]. It has been estimated that between 70-90 percent of medical visits are either directly or indirectly related to stress[10]. Stress has been defined as "physical or emotional cause of physical or mental tension", but the perception of stress can be just as debilitating as actual stress[11]. While many events are inherently stressful, we can use our mind-body connection to change how we react to our environment. Both stress or perceived stress can change our physiology by increasing our heart rate, blood pressure and blood sugar having a major effect on our immune system. In an acute situation these physiological changes help us escape danger, but chronic stress can cause hypertension[12], heart disease[13], diabetes[14] and cancer[15]. No one has control over our emotions except us. Appropriate reaction to our environment will change the way people both treat and react to us. Placing blame on others for our stress levels is counterproductive; blaming ourselves for getting sick is extremely harmful to not only our state of mind, but also our immune system. We are powerfully in control of how we react to our environment.

References

1. Morrow, J., et al., *Accuracy of measured and predicted residual lung volume on body density measurement.* Med Sci Sports Exerc, 1986. **18**(6): p. 647–652.
2. Smith, C., *A randomised comparative trial of yoga and relaxation to reduce stress and anxiety.* Complementary Therapies in Medicine, 2007. **15**: p. 77.
3. Moore, K. and A. Agur, *Essential Clinical Anatomy.* Third Edition, 2007. **Baltimore**(Lippincott Williams & Wilkins).
4. http://www.usatoday.com/news/health/2002-12-08-happy-main_x.htm., *What makes people happy?* 2010.
5. Haynes, R., *20 Questions to help you overcome barriers and find your passion.* The Wisdom Journal, 2011. http://www.thewisdomjournal.com/Blog/20-questions-overcome-barriers-find-your-passion/.
6. Roth, W., Wilhelm, F, Trabert, W, *Voluntary Breath Holding in Panic and Generalized Anxiety Disorders.* Psychosomatic Medicine, 1998. **60**: p. 671-679.
7. Paige, A., *Meditation Breathing Techniques.* eHow Health, 2011. http://www.ehow.com/how_4604742_meditation-breathing-techniques.html.
8. Kang, Y., *Mind-body approach in the area of preventive medicine: focusing on relaxation and meditation for stress management.* J Prev Med Public Health, 2010. **43**(5): p. 445-450.
9. Benson, H., Stuart, E, *The Wellness Book: The Comprehensive Guide to Maintaining Health and Treating Stress-Related Illness.* New York: Simon & Schuster, 1993.

10. Data, F.t.N.H.I.S., *Vital and Health Statistics Summary Health Statistics for U.S. Adults: National Health Interview Survey, 2010.* U.S. DEPARTMENT OF HEALTH AND HUMAN SERVICES, Centers for Disease Control and Prevention, National Center for Health Statistics, Hyattsville, Maryland, 2012. **Series 10**(No. 252): p. 11/2011.

11. Limm, H., Angerer, P, Heinmueller, M, Marten-Mittag, B, Nater, U, Guendel, H, *Self-perceived stress reactivity is an indicator of psychosocial impairment at the workplace.* BMC Public Health, 2010. **10**: p. 252.

12. Ron de Kloet, E., M. Joels, and F. Holsboer, *Stress and the brain: from adaptation to disease.* Nature Reviews Neuroscience, 2005. **6**(6): p. 463–475.

13. Iso, H., et al., *Percieved Mental Stress and mortality from cardiovascular diseae among Japanese men and women.* Circulation: Journal of the American Heart Assocation, 2002. **106**(1229-1236).

14. Mathews, E., Liebenberg, L, *A Practical Quantification of Blood Glucose Production due to High-level Chronic Stress.* Stress Health, 2011. **Dec 28**.

15. De Sousa, A., Sonavane, S, Mehta, J, *Psychological aspects of prostate cancer: a clinical review.* Prostate Cancer Prostatic Dis, 2012. **Jan 3**.

CHAPTER SIX
If I Get Sick What Are My Options?

Treating disease is most effective if done on every level: spiritually first, emotionally second, psychologically third, and physiologically fourth. By the time disease has manifested in our body, it has been developing in our mind and emotions for some time, and in our spirit long before that. Conventional health care treats at the final stage of disease – after it manifests physically. Amazingly once our mind and spirit have returned to a healthy state our body will energetically realign. This is known as spontaneous healing and is inexplicable in medical circles. Our mind is one of the most powerful tools we have to heal ourselves[1].

If you are currently ill it is crucial to follow a treatment plan, recognizing that there is more than one medical system available. Conventional medicine is effective at treating acute medical conditions however it is much less successful at treating chronic health problems. Alternative medical treatments can often relieve symptoms of disease without the side effects of pharmaceuticals. Complementary and Alternative Medicine (CAM) has many modalities that treat mind, body and spirit collectively. Common therapies include Ayurveda, acupressure, acupuncture, chiropractic and osteopathy, energy medicine, homeopathy, therapeutic massage, and Traditional Chinese Medicine (TCM). Integrating these treatments with conventional medicine can be very effective. There are definitely fraudulent practices being promoted as CAM, but there are also effective CAM treatments that people should be aware of and consider in their health plan. Integrative medicine combines the best of conventional medicine with complementary and alternative therapies. If you consider using CAM be sure to discuss these treatments with your health care practitioner. It is important to be educated regarding the different choices for CAM treatments available. I have initiated the research process by referencing articles published in peer reviewed medical journals. Some studies provide mechanisms for why these treatments work, devising sham treatments enabling better comparative trials, showing CAM treatments can lessen the side effects of many pharmaceuticals[2, 3]. Keep in mind that different CAM treatments work effectively in some people and not at all in others. Like pharmaceuticals, surgery and radiation CAM can be effective in different ways for different people.

Published Research for Common CAM therapies used today

Acupressure applies pressure to acupoints which are specific areas of the body believed to influence particular organs and their functions. Acupressure has been used longer than acupuncture (more than 2000 years) and works in a similar way, applying pressure instead of needles to relieve pain[4]. Acupressure has been reported to reduce tension and stress[5], prevent nausea[6], treat obesity[7], provide cardiovascular improvement after stroke[8], reduce cancer related fatigue[9], and improve cognitive brain function in traumatic brain injury[10].

Acupuncture involves inserting very fine needles at specific points on the body to stimulate and balance the flow of life energy known as *chi* or *qi*. Research has shown acupuncture to be beneficial in the treatment of allergic asthma[11], angina[12], anxiety [13, 14], insomnia[15, 16], carpal tunnel syndrome[17], chronic fatigue syndrome and fibromyalgia[18], Crohn's Disease[19], severe depression[20], depression during pregnancy[21], migraine headache[22], hypertension[23], incontinence[24], infertility in males due to low sperm count[25], infertility in women due to hormone disorders[26], reduction of inflammation[27], irritable bowel syndrome[28], chronic low back pain[29], neck and shoulder pain[30], Raynaud's Syndrome[31], schizophrenia[32], sciatica[33], and stroke rehabilitation[34].

Ayurveda is an ancient system of holistic healing originating in India that focuses on the balance between the physical, spiritual and psychological aspects of an individual. Ayurveda suggests that when a person is out of balance, illness, as well as emotional and spiritual negativity can result. Specific treatments are prescribed for detoxification, to release negative thoughts and emotions, and to promote rejuvenation. Therapies are individualized to the patient and include dietary changes, yoga, meditation, massage and herbal remedies. Ayurveda uses the Dosha constituents of Vata (ether, air), responsible for movement-related functions in the body such as respiration, circulation and thought; Pitta (fire, water), responsible for metabolism, hunger, thirst, digestion of food and assimilation of life experiences; and Kapha (water, earth), responsible for body structure to ascertain imbalances in the patient's spirit, mind and body. Ayurvedic treatments have been shown to reduce symptoms of fibromyalgia[35] and Rheumatoid arthritis[36]. It has also been effective in the treatment of cognitive disorders such as Alzheimer's disease[37]. Studies show that Ayurveda treatments are effective in treating stress[38], as well as obesity, metabolic syndrome, and diabetes mellitus[39]. [Caution: Although many Ayurvedic medical formulations are effective, take them with caution. A 2010 study published in the *Indian Journal of Clinical Biochemistry* reported three cases of lead toxicity associated with consumption of Ayurvedic medicines]. Work with an Ayurvedic Medical doctor. They are required to have many years of training in order to practice.

Chiropractic focuses on the relationship between the musculoskeletal system, the spine and nervous system, and how they relate to overall health. It is believed that the misalignment of the vertebrae in the spine can lead to health problems because it can block normal nerve function. Chiropractors use hands-on manipulation techniques to correct spinal misalignments and other issues. Beneficial results have been shown in patients with chest pain[40], balance issues associated with osteoarthritis[41], cyclic vomiting syndrome[42], migraine headaches[43], scoliosis[44] and neck pain[45].

Healing Touch is a standardized biofield therapy that uses gentle touch and movements in the patient's energy field with the goal of restoring energy and strengthening the patient's healing capacity[46]. It has been shown to significantly sustain a minimal decrease in natural killer cell cytotoxicity in patients undergoing radiation for cervical cancer as compared with relaxation training or usual care[47]. Healing touch has also been reported to reduce depression and improve quality of life in patients' with ovarian cancer[48], reduce pain, stress and anxiety[49], reduce cardiac surgery post-op symptoms[50], and improve cancer symptoms[51].

Homeopathy uses dilutions of substances from animals, plants and minerals to treat illness. It stems from the belief that "like cures like", meaning every substance that induces the symptoms of illness can treat that same illness when that substance is highly diluted. Homeopathy is used worldwide in dentistry and veterinary medicine. Homeopathy has been used to improve immunity[52], improve clotting in hemophiliacs[53], treat inner ear infections[54], improve symptoms of Rheumatoid arthritis[55], and allergies[56].

Massage therapy manipulates muscles, ligaments, tendons and other soft tissue by holding, moving, or applying pressure to an area of the body in order to relieve muscle tightness and spasms, increase blood flow, and promote the flow of lymph and other secretions in order to release toxic substances from the body. Massage therapy has been shown to ease pain and reduce side effects of chemotherapy and radiation in cancer patients[57], reduce pain and anxiety during labor[58], reduce pain and edema in sports injuries[59], improve pulmonary function in asthma patients[60], and be effective in reducing blood glucose levels in diabetic children[61]. It has also been shown to improve abnormal sensory responses and self-regulation of children with autism[62].

Osteopathy focuses on the musculoskeletal system and internal organs. Misalignment caused by muscle injuries, tension and poor posture is believed to block the free flow of blood and lymphatic fluids. Osteopaths use manipulative techniques to correct these misalignments to promote overall health benefits. Stress reduction and nutrition are an integral part of the treatment. Osteopathy has been shown to help low back pain[63, 64].

Polarity Therapy (PT) is an energy medicine system based on the idea that humans possess a biofield and the energy in their biofield can be manipulated for improved quality of life and increased health and wellness. Using touch, verbal interaction, exercise, nutrition and other methods[65], polarity practitioners seek to balance and restore the natural flow of energy in the biofield. PT has been reported to improve symptoms of cancer related fatigue[66] and dementia[67] and well as improve the quality of life for caregivers of oncology patients[68].

Touch Therapy (TT) is a four step process by which a practitioner focuses healing intention on the patient after "centering" (the practitioner's way of focusing), then "assessing" for imbalances in the biofield. Once the assessment is completed the practitioner facilitates the treatment and then performs a reassessment to integrate the treatment. TT has been shown to be effective in relieving tension headache pain for geriatric arthritic patients[69] and also significantly reduce neurological complications during bone marrow transplant[70]. TT has also been reported to decrease pain perception, reduce anxiety and enhance immune function in burn patients[71]. Because treatment is performed off the body this is especially beneficial for burn patients.

Pulsed Electromagnetic Field (PEMF) Therapy affects changes in the patient's biofield. PEMF has been approved by the Food and Drug Administration (FDA) as a therapeutic treatment of non-unions in patients with bone fractures[72-74]. Extremely low-frequency PEMF (around 15 Hz) induce voltages similar to those produced during normal mechanical deformation in connective tissue, has been reported to be most effective in eliciting osteogenesis (bone growth)[75] and healing fractures[76]. PEMF was also shown to be effective in the treatment of skin lymphoma[77], Alzheimer's Disease[78] and diabetic wound healing[79]. The 50 Hz, 10-55mT amplitude PEMF treatment has been reported to increase the lethal dose to cancer cells when used in combination with anticancer drugs[80].

Traditional Chinese Medicine (TCM) focuses on complex systems to prevent illness and support our body's natural capacity to heal itself. Treatment includes acupuncture, exercise, counseling, and herbal remedies. TCM takes into account the mind-body-spirit connection as well as effects of the home and social environment. Because every illness is a result of complex imbalances unique to each individual, treatment varies from patient to patient. Complex imbalances include yin (feminine) and yang (masculine) principles and their complementary sense organs, as well as tastes of bitter, sweet, pungent, salty and sour. TCM has been shown to reduce the symptoms of cancer and cancer related treatment[81, 82], chronic headache[83] and migraine headaches[84], has shown to be an analgesic and anti-inflammatory[85], and improve neural regeneration in patients after cerebral ischemic injury[86].

Yoga is an ancient Indian practice uniting the spirit, the body and mind accomplished through physical postures, controlled breathing exercises, and meditation, often accompanied by healthy lifestyle and search for higher consciousness. Yoga is not a religion but a philosophy and a way of life. Hatha yoga is most commonly practiced in North America and Europe, using a sequence of postures or asanas held statically or moved through dynamically in sequence, using the breath and hand positions for balance. Ashtanga yoga builds strength, stamina and flexibility, more commonly known in the United States as power yoga. Bikram yoga is practiced in rooms heated to 100 °F (39 °C). Profuse sweating loosens muscles and tendons while promoting inner cleansing. (This type of yoga should only be practiced after consulting with a physician). Research shows the practice of yoga can reduce pain, and increase energy, flexibility, and function during physical activity, as well as relieve stress and anxiety in breast cancer survivors[87]. It has also been shown to reduce pain associated disability[88], reduce stress[89], and as a complementary therapy for major psychiatric disorders[90].

Yoga Meditation combines the practice of yoga and meditation. This type of therapy is highly recommended for caregivers of terminal ill, geriatric or mentally ill patients. Caregivers are at risk for high stress levels, without relief or support, which can lead to many health problems. In a recent study published in the *International Journal of Geriatric Psychiatry* involving 49 caregivers who were taking care of family members with dementia, caregivers were asked to participate in a yoga meditation exercise. At the end of eight weeks 65 percent showed 50 percent improvement on a mental health scale. Among subjects who practiced breathing relaxation techniques, 31.2 percent showed reduction in symptoms of depression and 19 percent showed improvement on the mental health scale. More evidence was found on the cellular level, where the meditation group had a 43.3 percent improvement in telomerase activity, while the relaxation group saw only a 3.7 percent boost[91]. Telomeres are repetitions of DNA sequences at the end of a chromosome that protect it from damage which can lead to health problems. The higher our telomerase activity, the more durable our immune system will become[92].

Healing both mind and spirit can be intense, but fortunately it does not have to involve years of psychotherapy or spiritual counseling. There are various forms of energy medicine that have helped scores of people in minutes and days instead of weeks and years. **Energy psychology** (EP) addresses the relationship of energy systems to emotion, cognition behavior and health[93]. EP uses imagery, narrative, and hyperarousal associated with traumatic memory or threatening situations to resolve traumatic memory[94]. When the brain reprocesses traumatic memory, the new association is retained by reducing it to hyperarousal. This leads to treatment outcomes that involve less time with fewer repetitions and higher impact. These techniques show less chance of retraumatization[94]. During energy psychology treatments, mental/ emotional/spiritual problems are healed through our biofield. Our biofield is connected to our consciousness, our thoughts and our spirit, and includes the electrical activity of our nervous system, heart, meridians, biophotons, and chakras (explained in detail in chapter *From Conception to Birth*).

Energy Psychology treatments include *Thought Field Therapy (TFT)*[95], *Emotional Freedom Technique* (EFT)[96,97], Releasing Technique – Whole Health Easily and Effectively (WHEE)[98] and Tapas Acupressure Technique (TAT)[99]. Thought Field Therapy (TFT) has been shown to treat the fundamental causes of emotional disease and trauma by balancing the body's energy and eliminating most negative emotions within minutes. This technique has been reported to promote the body's own healing ability. A study using TFT with orphans of the 1994 Rwanda genocide reported outcomes that within minutes of being treated the orphans had an ability to return to normal life after months of depression[100]. By tapping on various body touch-points in a specific sequence, TFT stimulates the body's energy pathways similar to acupuncture, releasing stress, anxiety and emotions that are stored in the brain's thought field. Energy psychology has also helped victims of disaster relief[101], test anxiety[102], obesity[99], phobias[103] and fear[104], and Post Traumatic Stress Disorder (PTSD)[99].

Distortion in our biofield originating from mental and emotional imbalances occurs long before it manifests as disease in our physical body. Pharmaceuticals can be beneficial for acute illness. Over the counter Non-Steroidal Anti-Inflammatory Drugs (NSAIDs) relieve pain, antihistamines relieve runny nose and sinus problems, decongestants help us breathe better, but this temporary relief is treating the symptoms of the disease, not the cause. Dependence on pharmaceuticals is a very common problem in Western Medicine[105]. Many patients with chronic disease have multiple health problems. Common complaints are pain, stress, anxiety, depression, anger and angst, digestive disorders accompanied by allergies, poor sleep and fatigue. Cardiovascular disorders linked with hormonal imbalances and neurological problems contribute to multiple doctor visits where multiple tests are run and medications are taken for years if not lifetimes, causing dependence on drugs with multiple side effects[106]. Drug dependence begins when patients with chronic diseases seek new and more potent medications as current doses lose their effectiveness over time. Patients presenting multiple health problems take up to dozens of pills daily and continue to look for other treatments because their conditions do not improve. Taking a drug for condition A can cause conditions B, C, and D to become unbalanced. This phenomenon is known as side effects. Many patients need additional medication just to control side effects of the current drugs they are taking. Unfortunately, some have taken so many meds for so long that doctors cannot give them additional prescriptions when new conditions appear.

Disease is a signal that change is needed. When change does not happen, further insult occurs. Chronic illness leads to incurable disease and ultimately death. When we refuse to make changes in stress levels, eating habits, exercise practices, the soul is forced to leave the body in order to relieve itself of a poor quality of life. During times of change, prescriptions can be life saving, but changes must be made in order for disease to lessen or dissipate. For severe infection an antibiotic is a necessity. Surgery to remove cancers can be life saving. Combining orthodox treatment with alternative remedies is an option that has been shown to be more potent than either conventional or alternative treatments individually[107]. Combinations of western and eastern medicine significantly reduced the risk of death in stage I colon cancer by 95 percent, stage II colon cancer by 64 percent, stage III colon cancer by 29 percent, and stage IV colon cancer by 75 percent. There are additional treatments besides drugs and surgery which are healing. Choosing healthy, energetically balanced food is a must for sustaining wellness. Food can be very effective medicine (see chapter on *The Five Elements*). For the same reason we do not put sugar in our automobile fuel tank, we do not benefit from putting processed foods loaded with sugar and fat in our human fuel tanks. Healthy food sustains healthy beings. Movement and exercise are also medicines that move energy throughout our body, encouraging the life sustaining flow of *chi* or *prana*. Spending time in nature is medicine because through nature we reunite the energy of creation with our spirit.

How to Find Purpose in Life

What are our passions? Music (singing or song writing), art (dance, painting, sculpting), caring for others (physician, nursing, health care), construction, economics, or whatever peaks our interest. Passion is different than obsession. Passion brings joy and happiness; obsession leaves us feeling empty and depressed. Passion is fulfilling, obsession leaves us wanting more. Finding our passion and fulfilling our life's purpose is essential to health and healing. Begin with online research: http://www.lifehack.org/articles/lifestyle/how-to-find-your-passion.html

Read a book: *Wake Up...Live the Life You Love: Finding Life's Passion.* Journaling can help track emotional states and address reactions to stress. Changes are made through meditation, deep breathing, daily exercise and eating a healthy diet. It is also important to get emotional support for grief, despair and bullying.

There is one type of medicine that has no known side effects and treats all four levels of our being - it is known as energy medicine. Energy medicine (EM) is often described as a holistic health care system that works with the human energy field on the physical, mental, emotional and spiritual levels. It involves healing through physics instead of biochemistry. Cells and molecules respond to instruction given them by the biophysics of the body. Information travels in and around our body through our energy field. The information flowing through our biofield is harmonic and organized when our body is healthy, and it is disorganized and chaotic when our body is sick. Disease is a programmed state that occurs because our body, cells, molecules and atoms have been distorted by incoherent or chaotic messaging from unhealthy information in our environment. Energy medicine is non-invasive and can easily complement traditional forms of medical care for a complete self-care system. EM addresses physical illness as well as mental and emotional disorders. It also promotes high level wellness and peak performance[108]. Our bodies heal themselves by restoring and rebalancing energy that has become weak, disturbed or unbalanced. Healing is restored and maintained by reconnecting specific points along energy pathways using touch, pressure (acupoints) and direct stimulation of the biofield by the practitioners' hand. Energy medicine has been used to treat illness and relieve pain; stop the onset of disease as soon as it begins, stimulate immune function[47], relieve headache[109], release stress[49], improve memory[10], relieve arthritis[110], neck, shoulder and low back pain[111].

When it comes to health care we have choices. Whichever treatment option we choose, it must be our choice, not our doctor's, our spouse's, our children's or our parent's. Healing is personal. Disease begins on a personal level, healing returns in the same way.

References

1.   Goleman, D., Joel Gurin, J, *Mind Body Medicine: How to Use Your Mind for Better Health.* New York: Consumer Reports Books, 1993.

2.   Ausfeld-Hafter, B., Hoffmann, S, Seibold, F, Quattropani, C, Heer, P, Straumann *Status of alternative medicine in Crohn disease and ulcerative colitis patents: a questionnaire survey.* Forsch Komplementarmed Klass Naturheilkd, 2005. **12**(3): p. 134-138.

3.   Quimby, E., *The use of herbal therapies in pediatric oncology patients: treating symptoms of cancer and side effects of standard therapies.* J Pediatr Oncol Nurs, 2007. **24**(1): p. 35-40.

4.   Smith, C., Collins, C, Crowther, C, Levett, K, *Acupuncture or acupressure for pain management in labour.* Cochrane Database Syst Rev, 2011. **Jul 6;**(7): p. CD009232.

5.   Chang, K., Wong, T, Wong, T, Leung, A, Chung, J, *Effect of acupressure in treating urodynamic stress incontinence: a randomized controlled trial.* Am J Chin Med. , 2011. **39**(6): p. 1139-1159.

6.   Suh, E., *The effects of p6 acupressure and nurse-provided counseling on chemotherapy-induced nausea and vomiting in patients with breast cancer.* Oncol Nurs Forum, 2012. **39**(1): p. E1-9.

7.   Hsieh, C., Su, T, Fang, Y, Chou, P, *Effects of auricular acupressure on weight reduction and abdominal obesity in Asian young adults: a randomized controlled trial.* Am J Chin Med., 2011. **39**(3): p. 433-440.

8.   McFadden, K., Hernández, T, *Cardiovascular benefits of acupressure (Jin Shin) following stroke.* Complement Ther Med, 2010. **18**(1): p. 42-48.

9.   Zick, S., Alrawi, S, Merel, G, Burris, B, Sen, A, Litzinger, A, Harris, E, *Relaxation*

acupressure reduces persistent cancer-related fatigue. Evid Based Complement Alternat Med, 2011. **pii: 142913**(Sept 2).

10. McFadden, K., Healy, K, Dettmann, M, Kaye, J, Ito, T, Hernández, T, *Acupressure as a non-pharmacological intervention for traumatic brain injury (TBI).* J Neurotrauma., 2011. **28**(1): p. 21-34.

11. Joos, e.a., *Immunomodulatory effects of acupuncture in the treatment of allergic asthma: a randomized controlled study.* Altern Complement Med, 2000. **6**(6): p. 519-25.

12. Chao, e.a., *Nalaxone reverses inhibitory effect of electroacupuncture on sympathetic cardiovascular reflex responses.* American Journal of Physiology, 1999. **45**(6): p. H1227-H2134.

13. Chung, L.a., *The sedative effect of acupuncture.* Am J Chin Med 1979. **7**(3): p. 253-258.

14. Paraskeva, e.a., *Needling of the extra 1 point decreases BIS values and preoperative anxiety.* Am J Chin Med, 2004. **32**(5): p. 789-794.

15. Spence, D., et al *Acupuncture increases nocturnal melatonin secretion and reduces insomnia and anxiety: a preliminary report* J Neuropsychiatry Clin Neurosci, 2004 **16**(1): p. 19-28.

16. Sok, S., Erlen, J, Kim, K, *Effects of acupuncture therapy on insomnia.* J Adv Nurs, 2003. **44**(4): p. 375-384.

17. Naeser, B.a., *Carpal tunnel syndrome: clinical outcome after low-level laser acupuncture, microamps transcutaneous electrical nerve stimulation, and other alternative therapies.* J Altern Complement Med., 1999. **5**(1): p. 5-26.

18. Sprott, H., Franke, S, Kluge, H, Hein, G, *Acupuncture treatment of patients with fibromyalgia was associated with decreased pain levels and fewer positive tender points as measured by VAS (visual analog scale) and dolorimetry.* Rheumatology International, 1998. **18**(1): p. 35-36.

19. Joos, S., Brinkhaus, B, Maluche, C, Maupai, N, Kohnen, R, Kraehmer, N, Hahn, E, Schuppan, D, *Acupuncture and moxibustion in the treatment of active Crohn's disease: a randomized controlled study.* Digestion, 2004. **69**(3): p. 131-139.

20. Röschke, J., Wolf, C, Müller, M, Wagner, P, Mann, K, Grözinger, M, Bech S,, *The benefit from whole body acupuncture in major depression.* J Affect Disord, 2000. **57**(1-3): p. 73-81.

21. Manber, R., Schnyer, R, Allen, J, Rush, A, Blasey C, *Acupuncture: a promising treatment for depression during pregnancy.* J Affect Disord, 2004. **15**(83): p. 1.

22. Allais, G., De Lorenzo, C, Quirico, P, Lupi, G, Airola, G, Mana, O, Benedetto, C, *Non-pharmacological approaches to chronic headaches: transcutaneous electrical nerve stimulation, lasertherapy and acupuncture in transformed migraine treatment.* Neurol Sci 2003. **24**(Suppl 2): p. S138-S142.

23. Radzievsky, S., Lebedeva, O, Fisenko, L, Majskaja, S, *Function of myocardial contraction and relaxation in essential hypertension in dynamics of acupuncture therapy.* Am J Chin Med, 1989. **17**(3-4): p. 111-117.

24. Honjo, H., Kawauchi, A, Ukimura, O, Soh, J, Mizutani, Y, Miki, T, *Treatment of monosymptomatic nocturnal enuresis by acupuncture: a preliminary study.* Int J Urol, 2002. **9**(12): p. 672-676.

25. Siterman, S., Eltes, F, Wolfson, V, Lederman, H, Bartoov, B, *Does acupuncture treatment*

affect sperm density in males with very low sperm count: A pilot study. Andrologia 2000. **32**(1): p. 31-39.

26. Gerhard, I., Postneek, F, *Auricular acupuncture in the treatment of female infertility.* Gynecol Endocrinol., 1992. **6**(3): p. 171-181.

27. Woźniak, P., Stachowiak, G, Pieta-Dolińska, A, Oszukowski, P, *Anti-phlogistic and immunocompetent effects of acupuncture treatment in women suffering from chronic pelvic inflammatory diseases.* Am J Chin Med 2003. **31**(2): p. 315-320

28. Chan, J., Carr, I, Mayberry, J, *The role of acupuncture in the treatment of irritable bowel syndrome: a pilot study.* Hepatogastroenterology, 1997. **44** (1328-1330).

29. Carlsson, C., Sjölund, B, *Acupuncture for chronic low back pain: a randomized placebo-controlled study with long-term follow up.* Clin J Pain., 2001. **17**(4): p. 296-305.

30. He, D., Veiersted, K, Høstmark, A, Medbø, J, *Effect of acupuncture treatment on chronic neck and shoulder pain in sedentary female workers: a 6-month and 3-year follow-up study.* Pain, 2004. **109**(3): p. 299-307.

31. Appiah, R., Hiller, S, Caspary, L, Alexander, K, Creutzig, A, *Treatment of primary Raynaud's syndrome with traditional Chinese acupuncture.* J Intern Med, 1997. **241**(2): p. 119-124.

32. Chen, Z.a., *Comparison between electro-acupuncture with chlorpromazine and chlorpromazine alone in 60 schizophrenic patients.* Zhongguo Zhong Xi Yi Jie He Za Zhi, 1993. **13**(7): p. 408-409.

33. Duplan, B., Cabanel, G, Piton, J, Grauer, J, Phelip, X, *Acupuncture and sciatica in the acute phase. Double-blind study of 30 cases.* Sem Hop, 1983. **59**(45): p. 3109-3114.

34. Naeser, M., Alexander, M, Stiassny-Eder, D, Galler, V, Hobbs, J, Bachman, D, *Acupuncture in the treatment of paralysis in chronic and acute stroke patients -- improvement correlated with specific CT scan lesion sites.* Acupunct Electrother Res 1994. **19**(4): p. 227-249

35. Rasmussen, L., Mikkelsen, K, Haugen, M, Pripp, A, Fields, J, Førre, O, *Treatment of fibromyalgia at the Maharishi Ayurveda Health Centre in Norway II-a 24-month follow-up pilot study.* Clin Rheumatol, 2012. **Jan 27**.

36. Krishna, K., *The efficacy of Ayurvedic treatment for rheumatoid arthritis: Cross-sectional experiential profile of a longitudinal study.* Int J Ayurveda Res, 2011. **2**(1): p. 8-13.

37. Liu, L., Durairajan, S, Lu, J, Koo, I, Li, M, *In vitro screening on amyloid precursor protein modulation of plants used in Ayurvedic and Traditional Chinese medicine for memory improvement.* J Ethnopharmacol, 2011 **Sep 6**.

38. Basler, A., *Pilot study investigating the effects of Ayurvedic Abhyanga massage on subjective stress experience.* J Altern Complement Med., 2011. **17**(5): p. 435-440.

39. Sharma, H., Chandola, H, *Ayurvedic concept of obesity, metabolic syndrome, and diabetes mellitus.* J Altern Complement Med., 2011. **17**(6): p. 549-552.

40. Stochkendahl, M., Christensen, H, Vach, W, Høilund-Carlsen, P, Haghfelt, T, Hartvigsen, J, *Chiropractic Treatment vs Self-Management in Patients With Acute Chest Pain: A Randomized Controlled Trial of Patients Without Acute Coronary Syndrome.* J Manipulative Physiol Ther., 2012 **35**(1): p. 7-17.

41. Strunk, R., Hanses, M, *Chiropractic care of a 70-year-old female patient with hip osteoarthritis.* J Chiropr Med, 2011. **10**(1): p. 54-59.

42. Hubbard, T., Crisp, C, *Cessation of cyclic vomiting in a 7-year-old girl after upper cervical chiropractic care: a case report.* J Chiropr Med. , 2010. **9**(4): p. 179-183.
43. Chaibi, A., Tuchin, P, *Chiropractic spinal manipulative treatment of migraine headache of 40-year duration using Gonstead method: a case study.* J Chiropr Med., 2011 **10**(3): p. 189-193.
44. Morningstar, M., *Outcomes for adult scoliosis patients receiving chiropractic rehabilitation: a 24-month retrospective analysis.* J Chiropr Med. , 2011. **10**(3): p. 179-184.
45. Dunn, A., Green, B, Formolo, L, Chicoine, D, *Chiropractic management for veterans with neck pain: a retrospective study of clinical outcomes.* J Manipulative Physiol Ther., 2011. **34**(8): p. 533-538.
46. Menthe, J., Bulbrook MJ, *Healing Touch Level 1 Notebook (revised ed).* North Carolina Center for Healing Touch, 2002. **Carrboro, NC.**
47. Lutgendorf, S., Mullen-House, E, Russell, D, et al, *Preservation of immune function in cervical cancer patients during chemoradiation using a novel integrative approach.* Brain, Behavior, and Immunity, 2010. **24**: p. 1231-1240.
48. Judson, P., Dickson, E, Argenta, P, Xiong, Y, Geller, M, Carson, L, Ghebre, R, Jonson, A, Downs, L, *A prospective, randomized trial of integrative medicine for women with ovarian cancer.* Gynecol Oncol, 2011. **123**(2): p. 346-50.
49. Wardell, D., Weymouth, K *Review of studies of Healing Touch.* Journal of Nursing Scholarship: Image, 2004. **36**(2): p. 147-154.
50. Krucoff, M., *Healing touch, music, relaxation a plus for heart surgery patients.* The Lancet, 2005.
51. Post-White, J., Kinney, M, Savik, K, Gau, J, Wilcox, C, Lerner, I, *Therapeutic massage and healing touch improve symptoms in cancer.* Integr Cancer Ther., 2003. **2**(4): p. 332-344.
52. Coelho, C., de Fátima, J, Leal, M, Ferreira, E, Rezende, A, Imbeloni, A, Pereira, J, Smith, M, Burbano, R, de Assumpção, P, *Lymphocyte proliferation stimulated by activated Cebus apella macrophages treated with a complex homeopathic immune response modifiers.* Homeopathy, 2012. **101**(1): p. 74-79.
53. Kundu, T., Shaikh, A, Kutty, A, Nalvade, A, Kulkarni, S, Kulkarni, R, Ghosh, K, *Homeopathic medicines substantially reduce the need for clotting factor concentrates in haemophilia patients: results of a blinded placebo controlled cross over trial.* Homeopathy, 2012. **101**(1): p. 38-43.
54. Jacobs, J., *Homeopathy for acute otitis media-time for a definitive trial.* Homeopathy, 2012. **101**(1): p. 3.
55. Brien, S., Leydon, G, Lewith, G, *Homeopathy enables rheumatoid arthritis patients to cope with their chronic ill health: A qualitative study of patient's perceptions of the homeopathic consultation.* Patient Educ Couns, 2011. **Dec 14.**
56. Gründling, C., Schimetta, W, Frass, M, *Real-life effect of classical homeopathy in the treatment of allergies: A multicenter prospective observational study.* Wien Klin Wochenschr. , 2011. **Dec 8.**
57. Collinge, W., Macdonald, G, Walton, T, *Massage in supportive cancer care.* Semin Oncol Nurs, 2012. **28**(1): p. 45-54.
58. Mortazavi, S., Khaki, S, Moradi, R, Heidari, K, Rahimparvar, S, *Effects of massage*

therapy and presence of attendant on pain, anxiety and satisfaction during labor. Arch Gynecol Obstet, 2012. **Jan 21.**

59. Joseph, M., Taft, K, Moscwa, M, Denegar, C, *Deep Friction Massage for the Treatment of Tendinopathy: A systematic review of a classic treatment in the face of a new paradigm of understanding.* J Sport Rehabil, 2011. **Dec 30.**

60. Fattah, M., Hamdy, B, *Pulmonary functions of children with asthma improve following massage therapy.* J Altern Complement Med., 2011. **17**(11): p. 1065-1068.

61. Sajedi, F., Kashaninia, Z, Hoseinzadeh, S, Abedinipoor, A, *How effective is Swedish massage on blood glucose level in children with diabetes mellitus?* Acta Med Iran., 2011. **49**(9): p. 592-597.

62. Silva, L., Schalock, M, Gabrielsen, K, *Early intervention for autism with a parent-delivered Qigong massage program: a randomized controlled trial.* Am J Occup Ther., 2011. **65**(5): p. 550-559.

63. Licciardone, J., Brimhall, A, King, L, *Osteopathic manipulative treatment for low back pain: a systematic review and meta-analysis of randomized controlled trials.* BMC Musculoskelet Disord, 2005. **6**: p. 43.

64. Chou, R., Huffman, L *Nonpharmacologic therapies for acute and chronic low back pain: A review of the evidence for an American Pain Society/American College of Physicians clinical practice guideline.* Annals of internal medicine, 2007. **147**(7): p. 492–504.

65. Association, A.P.T., *APTA.* Standards for Practice, 2003. **Fourth Edition**: p. 2.

66. Mustian, K., Roscoe, J, Palesh, O, Sprod, L, Heckler, C, Peppone, L, Usuki, K, Ling, M, Brasacchio, R, Morrow, G, *Polarity Therapy for Cancer-Related Fatigue in Patients With Breast Cancer Receiving Radiation Therapy: A Randomized Controlled Pilot Study.* Integr Cancer Ther., 2011. **10**(1): p. 27-37.

67. Korn, L., Logsdon, R, Polissar, N, Gomez-Beloz, A, waters, T, Tyser, R, *A Randomized Trial of a CAM Therapy for Stress Reduction in American Indian and Alaskan Native Family Caregivers.* The Gerontologist, 2009. **32**: p. 1-10.

68. Pierce, B., *The use of biofield therapies in cancer care.* Clin J Oncol Nurs, 2007. **11**(2): p. 253-258.

69. Keller, E., Bzdek, V, *Effects of therapeutic touch on tension headache pain.* Nurs Res, 1986. **35**: p. 101-1016.

70. Smith, M., Reeder, F, Daniel, L. Baramee, J, Hagman, J, *Outcomes of touch therapies during bone marrow transplant.* Altern Ther Health Med, 2003. **9**: p. 40-49.

71. Turner, J., Clark, A, Gauthier, D, Williams, M, *The effect of therapeutic touch on pain and anxiety in burn patients.* J Adv Nurs, 1998. **28**: p. 10-20.

72. Holmes, G., *Treatment of delayed unions and nonunions of the proximal fifth metatarsal with pulsed electromagnetic fields.* Foot Ankle Int, 1994. **15**: p. 552-556.

73. Salzman, C., Lightfoot, A, Amendola, A, *PEMF as treatment for delayed helaing of foot and ankle arthrodesis.* Foot Ankle Int, 2004. **25**: p. 771-773.

74. Hock, D., *Electromagnetic field and magnets: investigational treatment for musculoskeletal disorders.* Rheumatic Dis Clinics North America 2000. **26**: p. 51-62.

75. McLeod, K., Rubin, C, *The effect of low-frequency electral fields on osteogenesis.* J Bone Joint Surg Am, 1992. **74**: p. 920-929.

76. Fredericks, D., Nepola, J, Baker, J, Abbott, J, Simon, B, *Effects of pulsed electromagnetic*

fields on bone healing in a rabbit tibial osteotomy model. J Orthop Trauma, 2000. **14**: p. 93-100.

77. Rodin, I., Lamotkin, I, Ushakov, A, Skvortsov, S, Velichko, A, Nekrasova, O, Lazarenko, E, *The use of a low-intensity eddy-current magnetic field in treating patients with skin lymphomas.* Voen Med Zh., 1996. **317**(12): p. 32-34.
78. Sandyk, R., *Alzheimer's Disease: Improvement of Visual Memory and Visuoconstructive Performance by Treatment with low intensity PEMF in Picotesla intensity.* International Journal of Neuroscience, 1994. **76**(3-4): p. 185.
79. Callaghan, M., et al., *Pulsed electromagnetic fields acclerate normal and diabetic wound healing by increasing endogenous FGF-2 release.* Plast Reconstr Surg. , 2007. **121**(1): p. 130-141.
80. Ruiz-Gomez, N., de la Pena, L, Prieto-Barica, M, Pastor, J, Gil, L. Matinez-Marillo, M, *Influence of 1 and 25 Hz, 1.5 mT magnetic fields on antitumor drug potency in a human adenocarcinoma cell line.* Bioelectromagnetics, 2002. **23**: p. 578-585.
81. Meng, Z., Kay, Garcia, M, Hu, C, Chiang, J, Chambers, M, Rosenthal, D, Peng, H, Wu, C, Zhao, Q, Zhao, G, Liu, L, Spelman, A, Lynn Palmer, J, Wei, Q, Cohen, L, *Sham-controlled, randomised, feasibility trial of acupuncture for prevention of radiation-induced xerostomia among patients with nasopharyngeal carcinoma.* Eur J Cancer, 2012. **Jan 27.**
82. Smith, M., Bauer-Wu, S, *Traditional chinese medicine for cancer-related symptoms.* Semin Oncol Nurs, 2012 **28**(1): p. 64-74.
83. Deng, G., *Chronic headache treated with acupuncture, Chinese massage and herbal medicine in Switzerland.* Zhongguo Zhen Jiu, 2011. **31**(12): p. 1121-1123.
84. Molsberger, A., *The role of acupuncture in the treatment of migraine.* CMAJ, 2012 **Jan 9.**
85. Yuan, H., Chen, R, Huang, D, Wang, Y, Wang, W, *Analgesic and anti-inflammatory effects of balance acupuncture on experimental scapulohumeral periarthritis in rabbits.* Zhongguo Zhen Jiu, 2011 **31**(12): p. 1106-1110.
86. Yang, Z., Chen P, Yu, H, Luo, WS, Wu, Y, Pi, M, Peng, J, Liu, Y, Zhang, S, Gou, Y, *Research advances in treatment of cerebral ischemic injury by acupuncture of conception and governor vessels to promote nerve regeneration.* Zhong Xi Yi Jie He Xue Bao, 2012 **10**(1): p. 19-24.
87. Galantino, M., Greene, L, Archetto, B, Baumgartner, M, Hassall, P, Murphy, J, Umstetter, J, Desai, K, *A qualitative exploration of the impact of yoga on breast cancer survivors with aromatase inhibitor-associated arthralgias.* Explore, 2012. **8**(1): p. 40-47.
88. Büssing, A., Ostermann, T, Lüdtke, R, Michalsen, A, *Effects of yoga interventions on pain and pain-associated disability: a meta-analysis.* J Pain., 2012. **13**(1): p. 1-9.
89. White, L., *Reducing stress in school-age girls through mindful yoga.* J Pediatr Health Care, 2012. **26**(1): p. 45-56.
90. Cabral, P., Meyer, H, Ames D, *Effectiveness of yoga therapy as a complementary treatment for major psychiatric disorders: a meta-analysis.* Prim Care Companion CNS Disord, 2011. **13**(4): p. 8.
91. Lavretsky, H., Siddarth, P, Nazarian, N, Khalsa, D, Lin, J, Blackburn, E, Irwin, M, *A pilot study of yogic meditation for family dementia caregivers with depressive symptoms: effects on mental health, cognition, and telomerase activity.* International Journal of Geriatric Psychiatry, 2012. **DOI: 10.1002/gps.3790**(March 11).

92. Eisenberg, D., *An evolutionary review of human telomere biology: the thrifty telomere hypothesis and notes on potential adaptive paternal effects*. Am J Hum Biol., 2011 **23**(2): p. 149-67.

93. Association, f.c.e.p., http://www.energypsych.org. 2012, Retrieved January 4, 2012.

94. Feinstein, D., *Energy psychology: a review of the preliminary evidence*. Psychotherapy: theory, research, practice, training, 2008a. **45**(2): p. 199-213.

95. Diepold, J., Goldstein, D, *Thought field therapy and QEEG changes in the treatment of trauma: a case study*. Traumatology, 2008. **5**: p. 1.

96. Brattberg, G., *Self-administered emotional freedom techniques (EFT) in individuals with fibromyalgia: a randomized trial*. Integrative Medicine: a clinicians journal, 2008. **august/september**.

97. Church, D., *The treatment of combat trauma in veterans using EFT (Emotional Freedom Techniques): a pilot protocol*. Traumatology, 2009. **15**: p. 1.

98. Benor, D., Ledger, K, Toussaint, L, Hett, G, Zaccaro, D, *Pilot study of emotional freedom technique (EFT), wholistic hybrid derived from EMDR and EFT (WHEE) and cognitive behavioral therapy (CBT) for treatment of test anxiety in university students*. Explore, 2009. **5**(6): p. 1.

99. Elder, C., Ritenbaugh, C, Mist, S, Aickin, M, Schnieder, J, Zwickey, H, Elmer, P, *Randomized trial of two mind-body interventions for weight-loss maintenance*. j of altern and complem med, 2007. **13**(1): p. 67-78.

100. Sakai, C., *Treatment of PTSD in Rwandan Child Genocide Survivors Using Thought Field Therapy*. Intl j of emerg mental health, 2006. **12**(1): p. 41-50.

101. Feinstein, D., *Energy psychology in disaster relief*. Traumatology, 2008b. **141**(1124-137).

102. Sezgin, N.O., B, Church, D, *The effect of two psychophysiological techniques on test anxiety in high school students*. International journal of healing and caring, 2009. **9**: p. 1.

103. Wells, S., Polglase, K, Andres, H, Carrington, P, Baker, A, *Evaluation of a meridian-based interventional EFT for reducing specific phobias of small animals*. Journal of Clinical Pharmacology, 2003. **59**(9): p. 943-966.

104. Waite, L., *Assessment of the emotional freedom technique: an alternative treatment for fear*. The Scientific Review of Mental Health Practice, 2003. **2**(1): p. 20-26.

105. Administration, U.S.D.O.H.A.H.S.-S.A.a.M.H.S., *Results from the 2010 National Survey on Drug Use and Health:Summary of National Findings*. Center for Behavioral Health Statistics and Quality, 2010.

106. Tsai, K., Chen, J, Wen, C, Kuo, H, Lu, I, Chiu, L, Wu, S, Chan, D, *Medication Adherence Among Geriatric Outpatients Prescribed Multiple Medications*. Am J Geriatr Pharmacother, 2012. **Jan 18**.

107. McCulloch, M., Broffman, M, van der Laan, M, Hubbard, A, Kushi, L, Abrams, D, Gao, J, Colford, J, *Colon Cancer Survival With Herbal Medicine and Vitamins Combined With Standard Therapy in a Whole-Systems Approach: Ten-Year Follow-up Data Analyzed With Marginal Structural Models and Propensity Score Methods*. Integr Cancer Ther., 2011. **10**(3): p. 240-259.

108. Oschman, J., *Energy Medicine in Therapeutics and Human Performance*. ISBN 10: 0750654007 2003. **Butterworth-Heinemann Medical**(ISBN-13: 9780750654005).

109. Foster, K., Liskin, J, Cen, S, Abbott, A, Armisen, V, Globe, D, Knox, L, Mitchell, M,

Shtir, C, Azen, S, *The Trager approach in the treatment of chronic headache: a pilot study.* Altern Ther Health Med., 2004. **10**(5): p. 40-46.

110. Brien, S., Leydon, G, Lewith, G, *Homeopathy enables rheumatoid arthritis patients to cope with their chronic ill health: A qualitative study of patient's perceptions of the homeopathic consultation.* Patient Educ Couns., Dec 2011. 89(3):507-16.

111. Marr, M., *The effects of the Bowen technique on hamstring flexibility over time: A randomized controlled trial.* Journal of Bodywork and Movement Therapies, 2011. **15**(3): p. 281 – 290.

CHAPTER SEVEN
What is Energy Medicine?

Energy medicine (EM) is medicine based on physics instead of biochemistry. Physics is the study of energy. Physics does not override biochemistry, it drives it. Both biology and chemistry behave according to the laws of physics. The human body is made of energy. It has structure (bones), plumbing (digestive tract), and electricity (nervous systems), all infused with energy. Matter is made of energy – cells, molecules, and atoms are all energy. Conventional medicine addresses disease with the same self assurance as a well seasoned auto mechanic, electrician or plumber. The patient is seen as equal to the sum of his parts instead of an integrated holistic organism. If we have plumbing problems, we go to the gastroenterologist. If we have joint, bone or skeletal problems, we go to the orthopedist. If we have neurological problems we see a neurologist. Energy medicine works with the electromagnetic signals in the body while pharmaceuticals affect chemical signals in the body. Healing with an integrated system that supervises the interaction of all these systems (neurological, skeletal, digestive, circulatory, endocrinal, integumentary, etc), is not only faster, it is more efficient. Such a system is known as the human energy field or human biofield. Most healing practitioners refer to this field as the human energy field, while scientists and the National Institutes of Health (NIH) refer to it as the human biofield.

Science has begun to measure this subtle but important energy field around our body and research is showing that when the natural flow of energy is obstructed, disordered, and depleted is when our body becomes diseased[1]. Energy medicine works with subtle forms of energy in and around our body, with the understanding that all illness results from disturbances in this field.

The continuous, uninterrupted flow of energy plays the main role in health and healing.

In 1989 the term *energy medicine* was coined by the International Society for the Study of Subtle Energy and Energy Medicine which studies the science of medical and therapeutic applications of subtle energies. Energy medicine came under government guidelines in 1992 when the National Institutes of Health (NIH) established the National Center for Complementary and Alternative Medicine[2]. According to the NIH, energy medicine is defined as a form of complementary and alternative medicine which has two distinct categories:

1. *Veritable* energy medicine, which uses mechanical vibration (sound) and electromagnetic radiation (light) in order to effect health and healing. It involves the use of specific, measurable wavelengths and frequencies to treat patients. Many of our body's electrical systems and electromagnetic fields are well known, and veritable forms of EM are being used in well established models for patients in today's medicine. Examples of veritable forms of EM are the use of lasers and magnetic pulses that have been found to be therapeutic. Commonly used forms of veritable EM such as electrocardiogram (EKG), electroencephalogram (EEG), Computerized Tomography (CT or CAT) Scan, Magnetic Resonance Imaging (MRI) and ultrasound equipment are currently being used in traditional medical applications.

Another veritable form of EM is Pulsed electromagnetic field (PEMF) therapy which is FDA-approved to fuse bones and has been cleared in certain devices to reduce swelling and joint pain[3]. PEMF has also been used to treat pain and edema in soft tissue for over 60 years. It has been firmly established that tissues such as blood, muscle, ligaments, bone and cartilage respond to biophysical input, including electrical and electromagnetic fields. Research shows that certain field strengths and frequencies of Pulsed Electromagnetic Field (PEMF) appear to be disease-modifying [Table 1].

Condition	B or Freq*	Treatment Duration	Treatment Number	1.2005 in
Alzheimer's[1]	5-8 Hz	30 min	2x	Significantly reduced cognitive dysfunction
Arthritis[2]	50 Hz	60 min	3x	Reduction of pain and inflammation
Back Pain[3]	64 Hz	16 min	until pain stops	Statistically significant potential for reducing pain
Carpal Tunnel Syndrome[4]	20 Hz	4 h	daily	Statistically significant short- and long- term pain reduction
Chronic Bronchitis[5]	30 mT	15-20 min	15x	Proved effective in patients suffering from chronic bronchitis when coupled with standard drug therapies
Cognitive Function[6]	900 MHz	2 h/week	55 weeks	Significantly reduced cognitive impairment in rats
Edema[7]	70 mT	15 - 30 min	6x	Significantly reduces acute edema
Fibromyalgia[8]	0.1-64 Hz	30 min	2x day/3 weeks	Improved function, pain, fatigue, and global status in FM patients
Gastroduodenitis[9]	100 Hz	6-10 min	8-10x	77 % of treatment patients experienced elimination of gastro-esophageal and duodenogastral refluxes compared to 29 % of controls
Glial Cells[10]	900 MHz	15 min	2-10 days	induces glial reactivity and biochemical modifications in the rat brain
Mastitis[11]	10-25 Hz	60 min	1x/2-3 mos	Significantly reduced post-op inflammation

Condition	B or Freq *	Treatment Duration	Treatment Number	1.2005 in
Multiple Sclerosis[12]	1-25 Hz	2-24h/day	Up to 5 weeks	PEMF device significantly alleviated symptoms
Migraine Headache[13]	27.12 Hz	1 h/day	5x/wk for 2 wks	Effective, short-term intervention for migraine, but not tension headaches
Nerve Regeneration[14]	2 Hz/ 0.3mT	1 h/day	10 days	Pre and post injury exposure suggests that PEMF influences regeneration indirectly
Neuritis[14]	10-100 Hz	6 min	10-12x	Produced beneficial effects in 93% of patients suffering from nerve problems
Oral Surgery Pre-Op[15]	5mT/30Hz	30 min	3-5 days prior to	Significantly reduce inflammation in clinical trials
Osteoarthritis[16]	25 G/ 5-24 Hz	25 G/5-24 Hz	18x	Rapid improvements of immuno-logical indices & alleviates symptoms
Pain and edema[17]	1mT or 5 mT	6 h/day	90 days	Significantly aids in clinical recovery
Post Traumatic Stress Disorder[18]	1Hz or 5Hz	40 sec or 8 sec/1 hr	20-30 days	Seventy-five percent of patients had a clinically significant antidepressant response
Rheumatoid Arthritis[19]	30 mT	30 min	15 – 20x	Reduces pain in chronic pain populations
Septic Shock[20]	50 Hz/ 2mT	6 h	1x	E. coli became more sensitive to antibiotics
Skin Ulcers[21]	75 Hz/ 2.7 mT	4 hr/day	for 3 months	Positive effects but only in small lesions
Tendonitis[22]	30 mT	60 min	10 – 20x	Significant beneficial effects

Condition	B or Freq *	Treatment Duration	Treatment Number	1.2005 in
Whiplash[23]	64 Hz	8 min	4x	Considerable and statistically significant pain reduction
Wound Healing[24]	245 mT	48 hours	8x/3-5 days	Postoperative pain was significantly reduced for a decrease in the need for analgesic resolve

* B=magnetic field; G=Gauss; T=Tesla; Hz=Hertz; 1 mT=10 Gauss

TABLE 1. Examples of different magnetic therapies which have been applied to treatment of inflammatory related conditions.

With veritable forms of EM we can adjust the vibrational frequency and field strength, to stimulate hormones[4], growth factors[5] and interleukins[6] with therapeutic frequencies affecting the way cells and molecules communicate and regenerate.

EM DEVICES

PEMF Knee Device

The PEMF knee device is an FDA-approved device consisting of a cuff that surrounds the knee. It has a coil and heating pads that send magnetic pulses and heat through the injured tissue. This device combines PEMF energy and thermal therapy in order to increase circulation, reduce swelling, relieve chronic pain and arthritis, as well as improve range of motion. Research has shown it benefits patients with osteoarthritis[3].

Transcutaneous Electrical Nerve Stimulation (TENS)

Transcutaneous electrical nerve stimulation (TENS) uses electric current to stimulate nerves to induce therapeutic treatment. These devices are usually connected to the skin using electrodes. A typical TENS device is able to modulate pulse width, frequency, and intensity of the electrical field it uses. TENS applied at frequencies above 50 Hz uses intensity below motor contraction (sensory intensity). TENS applied at frequencies below 10 Hz, use an intensity that produces motor contractions[7]. Studies show that TENS stimulates nerves in order to reduce both acute and chronic pain[8, 9].

PEMF Mats

These mats produce a therapeutic pulsed electromagnetic field (PEMF) that surrounds the entire body. PEMF delivered by whole-body mats are promoted in many countries for a wide range of therapeutic applications. Randomized, sham-controlled, double-blind trials focusing on osteoarthritis of the knee (3 trials) or the cervical spine (1 trial), fibromyalgia (1 trial), pain perception (2 trials), skin ulcer healing (1 trial), multiple sclerosis-related fatigue (2 trials), or heart rate variability and well-being (1 trial) have been performed, with outcomes varying between improvement and ineffective[10]. More research is needed to repeat outcomes. As of 2012 they have not been approved by the US Food and Drug Administration (FDA). PEMF mats are primarily advertised and distributed over the Internet, often used without medical supervision.

The second type of energy medicine is known as putative energy medicine.

2. *Putative* energy medicine is based on the idea that human beings are able to influence subtle forms of our body's energy with their hands, intentions, or meditation. By focusing on these subtle energies, EM practitioners are able to feel vibrational frequencies with their hands and align the human biofield with healing treatments[11]. Putative energy medicine is an all-inclusive term used for practices that include, but are not limited to Polarity Therapy (PT), Healing Touch (HT), Therapeutic Touch (TT), Bowen Technique, Reiki, Trager Approach, Qi gong, Traditional Chinese Medicine (TCM), acupuncture, homeopathy, electromagnetic (EM) therapy, energy psychology, Johrei, Touch for Health, Applied Kinesiology, Emotional Freedom Techniques (EFT), Nambudripad's Allergy Elimination Techniques (NAET), Tapa Acupressure Technique (TAT), Brennan Healing Science, Eden Energy Medicine, distance healing, intuitive medicine, and intercessory prayer.

The History of Energy Medicine

The earliest recorded use of a EM dates back to 2750 BC, when patients were exposed to shocks produced by electric eels[12]. The next recorded use was around 400 BC, when a Greek philosopher named Thales of Miletus rubbed amber and obtained static electricity[13]. The idea of an energy force field in and around our body has been an essential part of health and healing in various medical systems including Traditional Chinese Medicine, Indian Ayurvedic Medicine, Shamanic and Hermetic Medicine for thousands of years. The Egyptians, the Chinese, and the ancient Greeks all used magnetite or lodestone for healing[14]. In 1873 Edwin D. Babbitt observed energy fields around human bodies, but was ignored by scientific and medical communities until it was discovered that Babbitt was seeing neurocurrents flowing through the interhemispheric fibers of the corpus callosum in the brain[15]. EM suggests that when the body's energy is not flowing at optimal levels, it is due to weakened, concentrated or blocked energy resulting in mental, emotional and physical symptoms for the patient.

Yale University anatomy professor Harold Saxton Burr began studying the role of electricity in disease in 1929. In his book *Blueprint for Immortality: the Electric Patterns of Life,* he explains how he measured human energy fields with standard detectors. Modern research has confirmed the observations of Burr by concluding that both healthy and unhealthy people produce electrical charges and changes in magnetic fields around their body[1]. He argued that diseases appear in our energy field before symptoms of the disease appear in our body, suggesting that a disturbance in the field is a forewarning that illness is approaching. Burr felt a disruption of energy would precede a tumor. He theorized that if the disturbed energy field could be detected and restored to a normal harmonic state, the disease could be prevented.

In the late 1940s, Reinhold Voll began studying the electrical resistance of acupuncture pathways and comparing their conductance with specific diseases. Voll suggested that inflammation increases conductance, while organ degeneration and tissue necrosis decrease it [16, 17]. During the same era Hungarian physiologist and Nobel Prize winner Albert Szent-Györgyi began working on an idea that proteins in our body are semiconductors. A semiconductor is a substance that has the ability of being either a conductor or insulator, depending on the conditions, making it a good medium for the control of electrical current. Szent-Györgyi suggested that molecules in our body join together to form energy chains that guide electrons across long distances[18].

In the 1960s physicist Herbert Fröhlich studied how our energy field sets up vibrations that move around our body and radiate out into the environment. He showed how the body uses various electromagnetic frequencies to transmit different kinds of information from cell to cell explaining how these frequencies might be used to diagnose disease [19, 20]. Electromagnetic vibrations occur at many different frequencies, including visible and near visible light frequencies. Electromagnetic radiation (EMR) is characterized by its wavelength - the shorter the wavelength, the higher the frequency and the longer the wavelength, the lower the frequency. Fröhlich described frequencies in our body as highly organized and sensitive to information that is emitted by signals. These signals or vibrations penetrate the surface of a molecule, cell or organism and provide information that integrates processes for growth, injury repair, and bacterial and viral defense. Electromagnetic radiation is light having both particle and wave components[21]. An atom has both physical and energetic characteristics. The same can be said of biological cells. A cell-surface receptor, for example, stimulates a ligand to form a lock and key type connection to attract or repel atoms in the extracellular matrix of the receptor. This action creates change in the cell's intracellular molecules, initiating a signaling cascade that ultimately stimulates changes in gene expression. Electromagnetic signals effect cells, molecules and atoms throughout our body[22].

Other studies in the 1960s included work by electrical engineers Gerhard Baule and Richard McFee from Syracuse University who used a pair of run coils to detect a magnetic field produced by the electrical activity of a heart muscle[23]. Shortly afterwards Willem Einthoven discovered that heart electricity could be recorded with a galvanometer, and electrocardiograms (EKGs) began being used as a standard tool for medical diagnosis. A few years after Einthoven received the Nobel Prize for his discovery of heart electricity, Hans Berger showed that much smaller electrical fields could be recorded coming from the brain using electrodes attached to the scalp. This device is used to categorize health and disease of the brain and is known as the electroencephalograph (EEG)[24]. Today electroencephalograms record electrical activity along the scalp and measure voltage fluctuations resulting from ionic current flows in brain cells called neurons [25]. In the 1970s and 80s, orthopedic surgeon, Robert O. Becker conducted research in electrophysiology and energy medicine when he observed that properties of connective tissue (made of cells, fibers and extracellular matrix) surrounding the nervous system were operating on a direct current. Like Burr, Becker thought some sort of field surrounded the body, stimulating regeneration. He discovered that an electrostatic field could enable regeneration of a frog limb. He also reported that direct current (about 1 nanoampere) through a broken bone radically improves the growth and repair of bones[26]. Becker showed the part of the connective tissue layer called the perineurium sets up a low voltage current that controls injury repair. One of Becker's most important discoveries is that the perineurium (sheath surrounding nerve tissue) is sensitive to energy fields [27, 28] as Szent-Györgyi had suggested. Once it was discovered the body produces electrical and magnetic fields, it was then comprehensible how veritable EM devices measure and manipulate them.

Putative EM applies the same techniques as veritable EM using human hands instead of a device. Putative EM is based on the understanding that a therapist is able to facilitate healing by balancing disturbances in their clients' energy field. The practice of putative energy medicine dates back at least 5,000 years, as various healings are described in the Christian Bible and Jewish Torah (Acts 8:14-19; Genesis 27:27), along with other spiritual texts from cultures and religions around the world[29]. EM practitioners generate extremely low frequency (ELF) fields from their hands via meditation and intention[30, 31]. This energy entrains the biofield of the client and initiates a healing effect.

EM is carried out in an integrative and holistic manner. The concept of holism in medicine dates back to 460 B.C. with Hippocrates, the father of medicine, and is based on the idea that every aspect of our body and mind are interrelated to every other aspect of our being[32]. Manipulation of the holistic field allows for healing throughout the entire body, mind and spirit. EM influences the body's defense and repair mechanisms by manipulating and balancing vibrational frequencies in the body. This is done by entraining both the client and practitioner's breath, using it to move energy throughout the biofield. Practitioners ground themselves while working on clients so energy released from the client's field does not flow into the practitioner's field. Some practitioners are able to do this work without training however most practitioners learn their practice in a classroom setting.

How do Energy Medicine Techniques Work?

The hands of EM practitioners produce strong electromagnetic fields that affect our biofield in many ways. A measuring device called the Superconducting Quantum Interference Device (SQUID) is a magnetometer used to detect very weak biomagnetic fields. SQUID has detected frequencies coming from the hands of practitioners in the sub extra low frequency electromagnetic field (sub ELF-EMF) range of 0.3 to 30 Hz. The signal emitted by the practitioner is not steady or constant, but moves through the range of sub ELF-EMFs with the average range around 7 – 8 Hz.[33]. What I discovered while working on my PhD is that healing frequencies in the sub ELF-EMF range down regulate inflammatory responses such as tumor necrosis factor (TNF-a) and Nuclear Factor kappa B (NF-kB) which are cell proteins that play key roles in regulating immune responses [34, 35]. These proteins are affected by oscillations being produced via mechanical vibrations occurring in the polar membrane of cells[36]. EM techniques are capable of producing healing results because they can directly affect mechanical vibrations in both the membrane and cytoskeleton of human cells. Many research studies have detected the frequency limit of these cellular oscillations to be only 30 Hz[37-41], which is the same range of frequency coming from the hands of energy medicine practitioners. This range of frequencies affects not only the oscillations in our cells, but also gene expression during the cell signaling process[42]. A decrease in inflammation sets the stage for healing and regeneration.

Energy Medicine Therapies

Energy healing techniques such as Therapeutic Touch (TT) have found recognition in the nursing profession[43]. In 2005-2006, the North American Nursing Diagnosis Association approved the diagnosis of "energy field disturbance" in patients. Both Healing Touch (HT) and TT complement conventional health care and are used in collaboration with other approaches to health and healing. In acute care facilities it is used as an adjunct to other therapies, not in place of them. More than 30,000 nurses and doctors in hospitals use hands-on touch modalities each year and these procedures are documented as legitimate medical techniques[44]. The National Cancer Institute sponsors grant proposals involving Clinical Community Oncology Programs to study the effects of HT in reducing pain and suffering for end of life care[45]. The National Hospice and Palliative Care Organization recently published a Complementary Therapies in End-of-Life Care Manual with a full chapter on Healing Touch[46]. The AARP approved certified HT practitioners as providers for their American Whole Health Alternative Health & Wellness network[47]. HT certification workshops are taught in Schools of Nursing and Holistic education programs around the U.S.[48]. A best evidence synthesis of 62 published clinical trial articles studying proximally (practitioner and client in the same room) practiced biofield therapies such as biofield healing, energy healing, qi-therapy, polarity therapy, Reiki, therapeutic touch, and healing touch, suggests they are promising complementary interventions for reducing pain intensity in numerous populations, reducing anxiety for hospitalized populations, and reducing agitated behaviors in dementia, beyond what may be expected from standard treatment or nonspecific effects[49].

EM treatments, whether from the hands of a practitioner or a device, can interact with the biofield by the principle of superposition of waves in a non-linear dynamic system. As the treatment wave intersects the biofield of the client the net result is constructive interference. When the two individual waves are exactly in phase the result is large amplitude. This coherent (harmonic) field of the treatment overrides incoherent (noise) fields that are attributed to disease states.

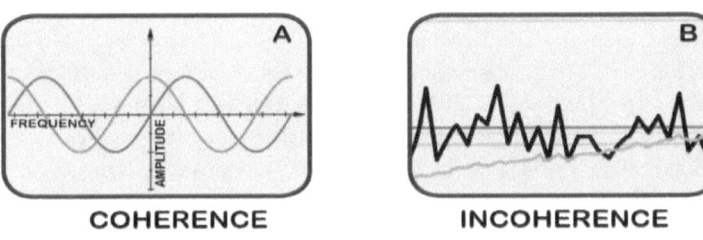

COHERENCE   **INCOHERENCE**

Figure 1A. Coherent (harmonic) field of treatment can override
Figure 1B. Incoherent fields through constructive interference.

Once they are in sync homeostasis is achieved. It is important to note this balanced state is not static. A balanced state of health depends on the capacity of a person to respond appropriately to stressors in their physical, mental or emotional environment. A balanced energy field is constantly adjusting to new stressors of internal and external environment (blood pressure, temperature, immune responses) moving them all toward dynamic equilibrium[50]. EM treatments interact with the biofield by the principle of superposition of waves in a non-linear dynamic system. As the treatment wave intersects the biofield of the client the net result is constructive interference. Studies suggest the application of energy medicine, whether from a medical device or from the hands of a practitioner, is a viable alternative or complement to conventional medicine. Free flowing energy throughout our body eliminates physical health problems attributed to pain, disease and structural dysfunction[51]. Energy medicine can greatly increase our energy level even if no specific problem exists. It is used as both a preventative and healing treatment.

References

1. Oschman, J., *Energy Medicine: The Scientific Basis*. Churchill Livingstone, 2000. **Edinburgh**.
2. NCCAM, http://nccam.nih.gov/health/whatiscam/. 2010.
3. Nelson, F., R. Zvirbulis, and A. Pilla, *The use of a specific pulsed electromagnetic field*

(PEMF) in treating early knee osteoarthritis Trans 56th Annual Orthopaedic Research Society Meeting, 2010. **New Orleans, LA**: p. 1034.

4. Schnoke, M., Midura, R, *Pulsed electromagnetic fields rapidly modulate intracellular signaling events in osteoblastic cells: comparison to parathyroid hormone and insulin.* J Orthop Res, 2007. **25**(7): p. 933-40.

5. Chalidis, B., Sachinis, N, Assiotis, A, Maccauro G, *Stimulation of bone formation and fracture healing with pulsed electromagnetic fields: biologic responses and clinical implications.* Int J Immunopathol Pharmacol, 2011. **24**(1 Suppl 2): p. 17-20.

6. Cossarizza, A., Monti, D, Bersani, F, Paganelli, R, Montagnani, G, Cadossi, R, Cantini, M, Franceschi, C, *Extremely low frequency pulsed electromagnetic fields increase interleukin-2 (IL-2) utilization and IL-2 receptor expression in mitogen-stimulated human lymphocytes from old subjects.* FEBS Lett., 1989. **248**(1-2): p. 141-4.

7. Robinson, A., Snyder-Mackler, L, *Clinical Electrophysiology: Electrotherapy and Electrophysiologic Testing (Third ed.).* Lippincott Williams & Wilkins, 2007.

8. Johnson, M., Martinson, M, *Efficacy of electrical nerve stimulation for chronic musculoskeletal pain: A meta-analysis of randomized controlled trials.* Pain, 2006. **130** (1): p. 157–165.

9. Dubinsky, R., Miyasaki, J, *Assessment: efficacy of transcutaneous electric nerve stimulation in the treatment of pain in neurologic disorders (an evidence-based review): report of the Therapeutics and Technology Assessment Subcommittee of the American Academy of Neurology.* Neurology, 2010. **74**(2): p. 173–176.

10. Hug, K., Röösli, M, *Therapeutic effects of whole-body devices applying pulsed electromagnetic fields (PEMF): A systematic literature review.* Bioelectromagnetics 2012. **33**(2): p. 95-105.

11. Benor, D., *Energy Medicine for the Internist.* Medical Clinics of North America, 2002. **86**(1): p. 105-125.

12. Kellaway, P., *The part played by electric fish in the early history of bioelectricity and electrotherapy.* Bulletin of the History of Medicine, 1946. **20**(112–132).

13. Babbitt, E., *The principles of light and color.* College of Fine Forces, 1873. **East Orange, NJ**.

14. Payne, B., *The body magnetic.* Privately published, 1990. **Santa Cruz, CA.**

15. Pansky, B., *Dynamic Anatomy and Physiology.* Macmillan, 1975. **New York**.

16. Voll, R., *The phenomenon of medicine testing in electro-acupuncture according to Voll.* Amer J Physiol Acupun, 1980. **8**(97–104).

17. Voll, R., *Twenty years of electroacupuncture diagnosis in Germany: a progress report.* Amer J Acupun, 1989(Special EAVedition): p. 5–14.

18. Szent-Gyorgyi, A., *Introduction to a submolecular biology.* Academic Press, 1960. **New York**.

19. Fröhlich, H. and F. Kremer, *Coherent Excitations in Biological Systems.* Springer-Verlag, 1983.

20. Fröhlich, H. and F. Kremer, *Biological Coherence and Response to External Stimuli.* Springer, 1988.

21. Pandey, P., *What is Electromagnetic Radiation.* http://www.buzzle.com/articles/what-is-electromagnetic-radiation.html, 2011.

22. Fels, D., *Cellular Communication through Light.* PLoS ONE, 2009. **4**(4): p. e5086.

23. Baule, G. and R. McFee, *Detection of the magnetic field of the heart.* Amer Heart Journ, 1963. **66**(95-96).

24. Berger, H., *Uber das elektrenkephalogramm des meschen.* Archiv fur Psykchiatria, 1929. **87**(527-570).

25. Niedermeyer, E. and F. da Silva, *Electroencephalography: Basic Principles, Clinical Applications, and Related Fields* Lippincot Williams & Wilkins, 2004 **Baltimore, MD.**

26. Becker, R. and J. Spadero, *Electrical stimulation of partial limb regeneration in mammals.* Bull, 1972. **NY Acad Med**(48): p. 627-641.

27. Becker, R., *The Body Electric: Electromagnetism and the Foundation of Life.* New York, 1985. **William Morrow & Co., Inc.**

28. Becker, R., *Cross Currents: the Perils of Electropollution, the promise of electro-medicine.* Jeremy P. Tarcher, 1990. **Los Angeles, CA.**

29. Dasa, A., *Tattva Prakasha.* Illuminations of Truth:, Retrieved 2-20-12. **2**(1): p. Srimad Bhagavatam 3.21.19.

30. Connor, M., G. Tau, and G. Schwartz, *Oscillation of Amplitude as Measured by an Extra Low Frequency Magnetic Field Meter as a Physical Measure of Intentionality.* Toward a Science of Consciousness, 2006.

31. Connor, M., M. Flores, and G. Schwartz, *The use of Triaxial ELF Magnetic Field Meter measurements as a predictor of capacity in Energy Medicine Practitioners in a research setting.* World Qi Gong Congress, 2004.

32. Hippocrates, *The Hippocratic Oath.* Translated by Michael North: National Library of Medicine, National Institutes of Health, 2012 retrieved(Hippocratic oath, nlm. nih.gov).

33. Zimmerman, J., *Laying on of hands healing and therapeutic touch: a testable theory.* J Bioelectromag Inst, 1990. **2**: p. 8-17.

34. Gilmore, T., *Introduction to NF-kB: players, pathways, perspectives.* Oncogene, 2006. **25**(51): p. 6680-6684.

35. Locksley, R.K., N, Leonardo, M, *The TNF and TNF receptor superfamilies: integrating mammalian biology.* Cell, 2001. **104**: p. 487-501.

36. Cifra, M., Fields, J, Farhadi, A, *Electromagnetic cellular interactions.* Progress in Biophysics and Molecular Biology, 2011. **105**: p. 223-246.

37. Korenstein, R., Levin, A, *Membrane fluctuations in erythrocytes are linked to mGATP-dependent dynamic assembly of the membrane skeleton.* Biophysical Journal, 1990. **60**: p. 773-737.

38. Tuvia, S., Almagor, A, Bitler, A, Levin, S. Korenstein R, Yedgar, S, *Cell membrane fluctuation are regulated by medium macroviscosity: evidence for a metabolic driving force.* Proceeding of National Academy of Science USA 94, 1997: p. 5045-5049.

39. Tuvia, S., Moses, A, Nathan, G, Levin, S, Korenstein, R, *β–adrenergic agonists regulate cell membrane fluctuations of human erythrocytes.* Journal of physiology, 1999. **516**(3): p. 781-792.

40. Popescu, G., Badizadegan, K, Dasari, R, Field, M, *Coherence properties of red blood cell membrane motions.* Journal of Biomedical Optics Letters 2006. **11**(4): p. 040503.

41. Popescu, G., Park, Y-K, Dasari, R, Badizadegan, K, Feld, M, *Coherence properties of red blood cell membrane motions.* Physical Review 2007. **E 76**(031902).

42. Ross, C., Harrison, B, *Regulation of A20 Gene Expression in Response to Pulsed Electromagnetic Field.* in publication, 2012.
43. Frisch, N., *Nursing as a Context for Alternative/Complementary Modalities.* Online Journal of Issues in Nursing, 2001. **6**(2): p. 2.
44. http://www.usatoday.com/news/health/2007-11-04-healing-touch, *Hands-on Touch.* 2010.
45. http://grants.nih.gov/grants/guide/rfa-files/RFA-AT-01-002.html, *Healing Touch for end of life care.*
46. http://www.nhpco.org, *Hospice and pallative care.* 2010.
47. AARP, http://www.aarp.org/, Retreived 2-20-12.
48. www.naturalhealers.com, *Natural Healers.* 2008.
49. Jain, S., Mills, P, *Biofield Therapies: Helpful or Full of Hype? A Best Evidence Synthesis.* International Journal of Behavioral Medicine, 2010. **17**(1): p. 1-16.
50. Rubik, B., *The Biofield Hypothesis: its biophysical basis and role in medicine.* J Altern Complem Medicine, 2002. **8**(6): p. 703-713.
51. Hsieh, L., et al., *Treatment of low back pain by acupressure and physical therapy: randomised controlled trial.* BMJ, 2006. **25**(332): p. 696-700.

Table References

1. Sandyk, R., *Alzheimer's Disease: improvement of visual memory and visuoconstructive performance treatment with picotesla range magnetic fields.* Int J Neurosci. , 1994. **76**(3-4): p. 185-225.
2. Wagner, T., Rushmore, J, Eden, U, Valero-Cabre, A, *Biophysical foundations underlying TMS: setting the stage for an effective use of neurostimulation in the cognitive neurosciences.* Cortex, 2009. **45**(9): p. 1025-1034.
3. Lee, P., et al., *Efficacy of pulsed electromagnetic therapy for chronic lower back pain: a randomized, double-blind, placebo-controlled study.* J Int Med Res, 2006. **34**(2): p. 160-167.
4. Weintraub, M. and S. Cole, *A Randomized Controlled Trial of the Effects of a Combination of Static and Dynamic Fields on Carpal Tunnel Syndrome.* Amer Acad Pain Med, 2008. **9**(5): p. 493-504.
5. Iurlov, V., T. Eksareva, and V. Dolodarenko, *The Efficacy of the Use of Low-Frequency Electromagnetic Fields in Chronic Bronchitis.* Voen Med Zh, 1989. **3**: p. 35-36.
6. Nittby, H., et al., *Cognitive impairment in rats after long-term exposure to GSM-900 mobile phone radiation.* Bioelectromagnetics, 2008. **29**(3): p. 219-232.
7. Morris, C. and T. Skalak, *Acute exposure to a moderate strength static magnetic field reduces edema formation in rats.* Am J Physiol Heart Circ Physiol, 2007. **294**: p. H50-H57.
8. Sutbeyaz, S., et al., *Low-frequency pulsed electromagnetic field therapy in fibromyalgia: a randomized, double-blind, sham-controlled clinical study.* Clin J Pain., 2009. **25**(8): p. 722-728.
9. Bukanovich, O., et al., *Sinusoidally-modulated currents in the therapy of chronic gastroduodenitis in children.* Von Kurortol Fizioter Lech Fiz Kult, 1996. **2**: p. 22-26.
10. Brillaud, E., A. Piotrowski, and R. de Seze, *Effect of an acute 900MHz GSM exposure on glia in the rat brain: a time-dependent study.* Toxicology., 2007. **238**(1): p. 23-33.

11. Navaratil, L., V. Hlavaty, and E. Landsingerova, *Possible Therapeutic Applications of Pulsed Magnetic Fields.* Cas Lek Cesk, 1993. **132**(9): p. 590-594.

12. Lappin, M., et al., *Effects of a pulsed electromagnetic therapy on multiple sclerosis fatigue and quality of life: a double-blind, placebo controlled trial.* Altern Ther Health Med, 2003. **9**(4): p. 38-48.

13. Sherman, R., N. Acosta, and L. Robso, *Treatment of migraine headaches with pulsing electromagnetic fields: A double blind, placebo controlled study.* Headache 1999. **39**(8): p. 567 - 575.

14. Sisken, B., *Nerve regeneration: implicaiton for clinincal application of electrical stimulation.* Paper presented at the 1st World Congress for Electricity and Magnetism in Biology and Medicine, 1992(Orlando, FL): p. June 14-19.

15. Hillier-Kolarov, V. and N. Pekaric-Nadj, *PEMF Therapy as an Additional Therapy for Oral Diseases.* EuroBioelectromag Assoc, 1992(1st Congress): p. 23-25.

16. Hulme, J., et al., *Electromagnetic Fields for the Treatment of Osteoarthritis.* Cochrane Review, Cochrane Library, Oxford, 2002. **1**(CD003523).

17. Dallari, D., *Effects of pulsed electromagnetic stimulation on patients undergoing hip revision prostheses: A randomized prospective double-blind study.* Bioelectromagnetics, 2009. **30**(6): p. 423-430.

18. Rosenberg, P., et al., *Repetitive Transcranial Magnetic Stimulation Treatment of Comorbid Posttraumatic Stress Disorder and Major Depression.* The Journal of Neuropsychiatry and Clinical Neurosciences 2002. **14**: p. 270-276.

19. Shupak, N., *Therapeutic Uses of Pulsed Magnetic Field Exposure.* Radio Science Bulletin, 2003. **307** (December): p. 9-32.

20. Gaafar, E., et al., *Stimulation and control of E. coli by using an extremely low-frequency magnetic field.* Romanian J. Biophys, 2006. **16**(4): p. 283-296.

21. Jeran, M., *PEMF Stimulation of Skin Ulcers of Venous Origina in Humans: Preliminary Report of a Double Blind Study.* Bioelectromagnetics, 1987. **6**(2): p. 181-188.

22. Binder, A., *Pulsed electromagnetic field therapy of persistent rotator cuff tendinitis.* Lancet, 1984. **8379**: p. 695-698.

23. Thuilea, C. and M. Walzlb, *Evaluation of electromagnetic fields in the treatment of pain in patients with lumbar radiculopathy or the whiplash syndrome.* NeuroRehabilitation, 2002. **17** (1): p. 63–67.

24. Man, D., *Effect of Permanent Magnetic Field on Postoperative Pain and Wound Healing in Plastic Surgery.* . Second World Congress for Electricity and Magnetism in Biology and Medicine, 1997.

CHAPTER EIGHT
The Human Biofield

The human biofield is a complex combination of overlapping energy patterns that characterizes the unique spiritual, mental, emotional and physical makeup of an individual. (see figure 1).

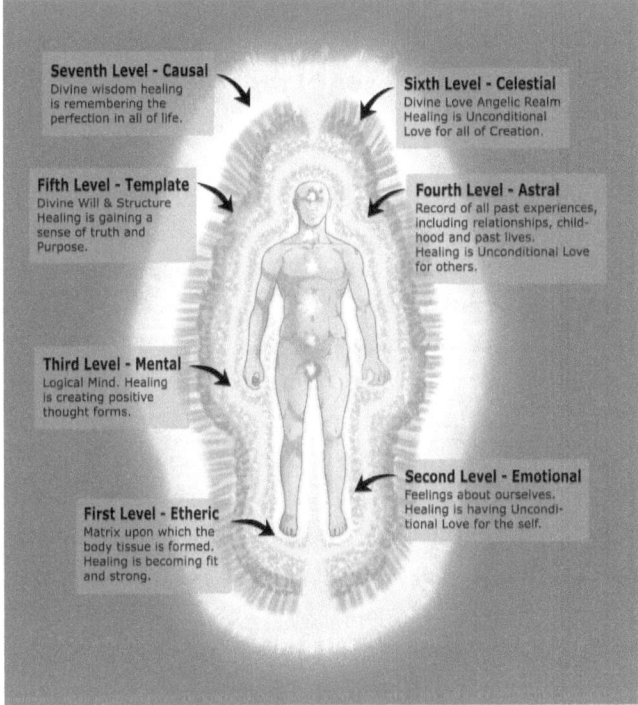

Figure 1. The field surrounds the outer physical body in all directions and extends outward both internally and externally

The first level of the human energy field, known as the etheric field, is closest to the body. It provides the matrix through which energy forms matter. Energy imbalances start at the spiritual level of the causal and celestial body then step down from mental and emotional states in the second and third level matrices of the field. The causal level of the human energy field is the beginning of all human experience. Original disease originates when there is separation or lack of integration between the sixth and seventh levels of the energy field, where energy no longer flows between the physical (self/ego), the mental self (mind) and the spiritual self (soul). Our ability to immerse our self into human life, and to seek and perform our soul's purpose in human form, begins on the etheric level. The second level of the biofield is the emotional level. At this level we learn to accept or ignore our emotions. Acceptance of our emotions and feelings about self and others, along with our ability to love unconditionally, are determining factors in whether disease manifests physically. The third level of our energy field reflects our mental state. Whether we possess positive healing thoughts that increase our health and well being or damaging negative thoughts that decrease the free flow of energy, we create an environment for healing or one for promoting disease. The fourth level of the human biofield is the astral plane. The astral plane is connected to the multi-dimensional realms of past lives and past experience. At this level the act of forgiveness creates a healing response. The astral plane is where grudges, hurt, revenge, pain, release, forgiveness, peace and freedom are held. Judgmental thoughts towards self and others produce low frequency energy in our biofield. It has been said that holding a grudge is like drinking poison and hoping it kills the person we are angry with. If this level is not cleared of low frequency energy on a regular basis, it becomes a breeding ground for disease. The fifth level of the human energy field connects with our divine will. It functions as guidance for our life's purpose – our reason for living. Energetic separation between the fifth level template and our mental and emotional state will block the energy flow between our mind and our soul. Here is where depression begins. When depressed we lose our love of life and our love of self. At the sixth level of our biofield we find our connection to creation. In order to stay connected to both heaven and earth it is imperative we spend time in nature, be near trees and rocks, and commune with the earth. The sixth level of our field is where we have "mountain top experiences" and are able to see love and beauty in all things. At this level we feel deep unconditional love for life - humans, plants and animals. Stagnant or blocked energy at the sixth level causes feelings of loneliness, isolation and we become incapable of receiving love and affection. The seventh level of the human energy field is our connection with the divine. Divine energy is necessary in order for life to exist - without it we would die. Connecting with the divine through meditation or prayer empowers us, energizes us, it fuels our chakras and meridians to provide life force energy throughout our body. At the seventh level of the human energy field all healing begins. The human energy field is a step down process by which higher frequencies systematically decrease to lower frequencies with each transformation enhancing or blocking the experience between our soul and our body. Our mind is the information processor of our human experience, while our emotions are the reflections of our soul's desire. These multiple levels have been understood for thousands of years. They are an integral part of Indian Ayurveda and Traditional Chinese Medicines. Shamans too understand the intricacies of human biofields.

Disease manifests in the physical body as a process that steps down through different chakras, through meridians into specific areas of the body. Disease evolves when our soul feels separated from the divine or has lost its purpose for having an earthly experience. Once our soul's mission has been forgotten, its purpose for existing in human form is no longer relevant. Our soul becomes subdued and our mind shifts its focus on sustaining the body - a matter of seeking food, clothing and shelter. This superficial form of existence is the reason why money does not make people happy. Money can provide comfort but not happiness. Once the soul's reason for existing in is no longer evident, the soul looks for a way out of the body. It longs to go home – to return to the divine or universal life force. It amazes me how many people begin to live only after they have been diagnosed with disease. Through their disease they see the purpose of living a joy filled vibrant life. Disease can be a gift!

Medical science uses a biochemical model of molecules, tissues and organs to focus on an organized structure-function relationship of health and disease. This model needs to be expanded to deeper levels in order to include electromagnetic and quantum processes that play a major role in how nature organizes itself. Quantum physics teaches us there is no difference between energy and matter, stating that all systems in an organism from the atomic to the molecular level are constantly in motion; that resonance is important in understanding how electromagnetism can have different effects on the body; and that all things resonate - but at different vibrational frequencies. Each type of body tissue resonates differently and emits characteristic signals from the nuclei of its atoms[1]. Energy in living tissues is transferred by excited electrons moving within semiconducting matrices[2]. In addition to being a protective shield, the cell membrane is also a powerful signal amplifier that resonates and governs the flow of ions into and out of the cell through minute changes in electrical tension which can be affected by exogenous fields. This model needs to be expanded to deeper levels in order to include electromagnetic and quantum processes which play a major role in how nature organizes itself. The cell membrane is often considered the main target for magnetic field signals. Results point to the Hall voltage as the catalyst of semiconduction, rather than ionic currents[3].

Because the human biofield is electromagnetic in nature it reacts to other electromagnetic fields. Russian scientists have measured a holographic type of bioenergy using an optical imaging device. Research suggests the existence of subtle radiation linked to DNA supporting the hypothesis of an intact energy field[4]. This DNA radiation effect was first detected at the Russian Academy of Sciences in Moscow during experiments measuring the vibrational modes of DNA in solution using a sophisticated laser photon correlation spectrometer[5]. Experiments showed that when DNA was removed from a scattering chamber and measured, it looked very different from the measurements taken from controls (before the DNA was placed in the chamber). Multiple replications with recalibrated equipment showed a new field was being stimulated from the physical vacuum. They called this new field the "DNA phantom" to suggest that its origin was related, but not physically linked, to the measured DNA. What makes this discovery different from numerous other attempts to measure and identify human biofields is the DNA phantom field has the ability to be coupled to conventional electromagnetic fields of laser radiation and can be consistently detected and positively identified using standard optical techniques.

In 1939, a Soviet electrician named Semyon Davidovich Kirlian developed a technique to photograph unique phenomena on human subjects. The photographic technique, known as Kirlian photography, showed people having illnesses which had not been diagnosed prior to the photographs. A similar technique is being used today called Digital Infrared Thermal Imaging (DITI). DITI detects angiogenesis in a tumor as small as the head of a pin. Angiogenesis is the beginning of blood vessel growth. Tumors receive nutrients and oxygen through blood supplies via angiogenesis proceeding tumor growth. DITI is used as an adjunct to mammorgraphy for early breast cancer detection and monitoring of abnormal physiology. It can establish risk factors for the development or existence of disease. It is a non-invasive, non-contact device using no radiation, to diagnose the onset of rheumatology, neurology, extracranial vessel disease, neuromusculoskeletal disorders, vertebrae (nerve problems/arthritis), oncology (skin cancer), and sports injuries[6-8]. It is also used for early detection of lower extremity vessel disease and tissue viability in diabetes[9] and Grave's disease[10].

Magnetic encephalographs (MEGs) use probes outside the body (around the head area) to read biofield activity without touching the body. Studies show brain signals broadcast fields[11] which are intelligent, meaning they hold information. The human energy field (or biofield) is described as a complex dynamic of electromagnetic fields that include individual oscillating electrically charged moving particles such as ions, biophotons, and molecules, which ultimately create standing waves[12], thus the biofield can be described in terms of the wave-particle duality. It is in this field where information is exchanged. Quantum physics explains how the character of matter is ultimately determined by the field[13]. Biophysicists study the fundamental processes of life by applying the methods of physics and chemistry to biological systems. We are beginning to understand how light (electromagnetic radiation) works within the molecular and atomic realms. When atoms or molecules absorb light, this light energy excites quantized particles to higher energy levels. The excitation depends on the wavelength of the light. Both light and quantum electrodynamics (QED) are becoming more important in the field of biology[14]. Fritz-Albert Popp discovered that living cells in the human body and in other organisms emit ultra-weak light[15]. Popp's discovery is important in understanding the energy mechanisms in the function of all living things. Energy medicine practitioners understand every organ in the body has its own energy field and the physical body can be affected by emotions, mental thought and intentions, as well as extraneous electromagnetic fields it comes in contact with. The vibrations of the human energy field are believed to emanate from the atomic and subatomic molecular activity of human cells[16]. Quantum physics suggests all matter contains potential energy which is released in small quantities as electrons spin off. An interesting characteristic of energy emission from any object is that it stays somewhat organized in its fields. It has a tendency to remain stable and does not randomly dissipate[17]. Living substances emit a higher frequency with more dynamic pattern changes than inert substances. Biofield vibrations are like tuning forks, acting as both transmitters and receivers of sound. They resonate at specific harmonic pitches when we are healthy. When we are not healthy a non-coherent type noise vibrates from our cells and our biofield[18]. This noise can be manipulated to vibrate at healthy harmonic frequencies using EM treatments.

The human biofield both permeates the entire body and radiates to the outside of our body's surface. Its outer shape appears oval and extends about two feet beyond the physical body; however, this shape can be extended even further or contracted closer to the physical body depending on the health and emotions of the individual. For example, when a person is feeling emotions of unconditional love, the aura can expand to several feet and radiate bright hues of gold or white. But if the same person is feeling threatened physically or emotionally, the entire aura may collapse to a much denser pattern within only a few inches of the body.

Disturbances in the coherence of energy patterns in the field are indicative of diseases which are about to develop. In order to correct field disturbances, it is important to understand how the energy field is constructed. There are three basic ways of describing Energy Field structure - in terms of energy meridians, chakras and energy bodies. Energy meridians are the internal energy pathways throughout the physical body that energetically connect a person's organs and their sub-systems (e.g., circulatory, respiratory, skeletal, endocrine, nervous, digestive, etc.). The chakra system within the human body consists of seven major chakras and many minor chakras. To those who can see energy fields, a major chakra resembles a spinning wheel when looking directly into the chakra; however, when viewed from the side, it looks more like an energy vortex resembling the shape of a tornado (See figure 2).

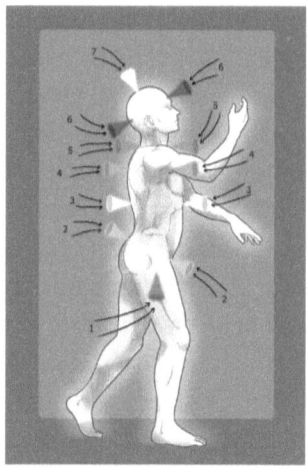

Figure 2. Chakra energy funnels are tight and compact near the surface of the skin and gradually widen as they extend outside the physical body to the outer edge of the aura. Energy bodies consist of seven different distinct layers—etheric (1), emotional (2), astral (3), mental (4), template (5), celestial (6), and causal (7).

The physical body is also considered an energy body since all matter is made up of energy. The higher subtle energy bodies overlap and penetrate the complete physical body the same way different radio, television, cell phone and satellite signals overlap each other in space but can be individually identified by a specific frequency. When an EM practitioner places her hands on a client, the healing frequencies are not only sent to the physical body, but to each higher energetic body. Therapeutic frequencies emitted from the practitioner promote healing not only at the physical level, but also in the etheric, emotional, mental and spiritual levels simultaneously.

The Human Energy Field has been validated in scientific laboratories as photon (light) emissions using photometers and color filters. As Valerie Hunt describes in her book, *Infinite Mind*, "During a research study that harnessed the complex wave forms and resonance that connects us to our Source, we found the human energy field vibrations were as much as 1000 times higher in frequency than the electrical signals of nerve and muscle, with continuous, dynamic modulation - unlike the pulsating signals of the nervous system"[16]. High frequency vibrations are indicative of a higher resonance emitted from a field above the resonance of the physical body alone.

The human energy field is electromagnetic in nature. It is a system of coherent energy patterns and light made of photon emissions, creating both wave and particle matter. It is a system of transmitters and receivers of vibration that program healing information into our body through its environment. Our biofield governs the physical matter of our body. Disease is scrambled or incoherent information in our field. Healing occurs by altering the distortion of information and applying coherent, harmonic resonances that naturally occur in a healthy body. Correcting the incoherence of information is the key to healing disease.

References

1. Farran, A., Teller, S, Jia, F, Clifton, R, Duncan, R, Jia, X, *Design and characterization of a dynamic vibrational culture system.* J Tissue Eng Regen Med, 2011. **Nov 18. doi: 10.1002/term.514. [Epub ahead of print]**.
2. Szent-Gyorgyi, A., *Introduction to a submolecular biology.* Academic Press, 1960. **New York**.
3. Funk, R. and T. Monsees, *Effect of electromagnetic fields on cells: physiological and therapeutical approaches and molecular mechanisms of interaction.* Cell, Tissues, Organs, 2006. **182**(2): p. 59-78.
4. Gariaev, P., Grigor'ev, K, Vasil'ev, A, Poponin, V, Shcheglov, V, *Investigation of the*

Fluctuation Dynamics of DNA Solutions by Laser Correlation Spectroscopy. Bulletin of the Lebedev Physics Institute, 1992. **11-12**: p. 23-30.

5. Gariaev, P., Poponin, V, *Vacuum DNA phantom effect in vitro and its possible rational explanation.* Nanobiology, 1995. **in press**.

6. Herman, C., Cetingul, M, *Quantitative visualization and detection of skin cancer using dynamic thermal imaging.* J Vis Exp. , 2011. **5**(51): p. 2679.

7. Kontos, M., Wilson, R, Fentiman, I, *Digital infrared thermal imaging (DITI) of breast lesions: sensitivity and specificity of detection of primary breast cancers.* Clin Radiol, 2011. **66**(6): p. 536-539.

8. Lee, J., Lee, J, Song, S, Lee, H, Lee, K, Yoon, Y, *Detection of suspicious pain regions on a digital infrared thermal image using the multimodal function optimization.* Conf Proc IEEE Eng Med Biol Soc, 2008: p. 4055-4058.

9. Ring, F., *Thermal imaging today and its relevance to diabetes.* J Diabetes Sci Technol. , 2010 **4**(4): p. 857-862.

10. Shih, S., Lin, M, Li, H, Yang, H, Hsiao, Y, Chang, M, Chen, C, Chang, T, *Observing pretibial myxedema in patients with Graves' disease using digital infrared thermal imaging and high-resolution ultrasonography: for better records, early detection, and further investigation.* Eur J Endocrinol, 2011. **164**(4): p. 605-611.

11. McCraty, R., Atkinson, M, Tomasino, D, Bradley, R, *The Coherent Heart: Heart-Brain Interactions, Psychophysiological Coherence, and the Emergence of System-Wide Order* HeartMath Research Center, Institute of HeartMath, 2006. **Publication No. 06-022**(Boulder Creek, CA).

12. Rubik, B., *The Biofield Hypothesis: its biophysical basis and role in medicine.* J Altern Complem Medicine, 2002. **8**(6): p. 703-713.

13. Jefimenko, O., *Presenting electromagnetic theory in accordance with the principle of causality.* Eur. J. Phys, 2004. **25**: p. 287.

14. Feynman, R., *The Strange Theory of Light and Matter.* Princeton University Press, 1985.

15. McTaggart, L., *The Field: the quest for the secret force of the universe.* Harper Collins Publishers, 2002. **New York**.

16. Hunt, V., *Infinite Mind: Science of the Human Vibrations of Consciousness.* Malibu Publishing Company, 1996. **Malibu, CA**.

17. Dirac, P., *The Principles of Quantum Mechanics.* Clarendon Press, Oxford, 1930.

18. Hunt, V., *The promise of bioenergy fields: an end to all disease.* http://www.spiritofmaat.com/archive/nov1/vh.htm, 2012 retrieved May 15.

CHAPTER NINE
The Five Elements and How they Relate to Health and Healing

Ayurveda is the Hindu system which follows the philosophy that there are five elements in nature and in the body. These five elements are ether, air, fire, water, and earth. Every part of the human body is the manifestation of one or more of these elements[1]. For example, air is manifested as the beating of the heart and in the expansion and contraction of the lungs. Fire is manifested in the digestive tract, as body temperature, and metabolism. Ayurveda understands that staying in harmony with nature is essential for optimum health. Written in India around 900 BC, the Ayurveda (which means "Science of Life"), combines descriptions of disease with information on herbs and healing treatments. Ayurvedic medicine is one of the world's oldest medical systems, originating in India and time tested for thousands of years.

In Ayurveda *doshas* control the activities of the body and determine a person's chances of developing certain types of diseases. Disease is believed to be related to the way doshas are balanced through the state of our physical and mental body as well as our lifestyle. Imbalances can be caused by a person's age, unhealthy lifestyle, or diet; too much or too little mental and physical exercise; the seasons or inadequate protection from the weather; and exposure to chemicals and germs. Each dosha is made up of two of the five basic elements: vata (ether and air), pitta (fire and water), and kapha (water and earth). Each person has a unique combination of the three doshas, although one dosha is usually prominent. The five elements are constantly being formed and reformed by food, activity, and bodily processes. An imbalance in any or all of the five elements will produce symptoms that are unique to individual people.

The *vata* dosha combines the elements of ether and air. It is considered the most powerful dosha because it controls very basic body processes such as cell division, the heart, breathing, discharge of waste, and the mind. Vata can be aggravated by fear, grief, lack of sleep, or eating before the previous meal is digested. People with vata as their main dosha are thought to be especially susceptible to skin and neurological conditions, rheumatoid arthritis, heart disease, anxiety, and insomnia.

The *Pitta* dosha represents the elements of fire and water. Pitta controls hormones and the digestive system. A person with pitta imbalance may experience low frequency emotions such as anger and may have physical symptoms such as heartburn within two to three hours of eating. People with a predominantly pitta constitution are thought to be susceptible to hypertension, heart disease, infectious diseases, and digestive conditions such as Crohn's disease.

The *kapha* dosha combines the elements of water and earth. Kapha is aggravated by eating too many sweets, eating and drinking foods with too much salt and water (especially in the springtime), greed, and sleeping during the daytime. Those with predominant kapha are thought to be vulnerable to diabetes, cancer, obesity, and respiratory illnesses such as asthma. Ayurvedic treatment is tailored to each person's constitution. Practitioners expect patients to be active participants in their own health care because many Ayurvedic treatments require changes in diet, lifestyle, and exercise habits. Ayurvedic treatments include eliminating impurities, reducing symptoms, boosting immunity, reducing worry and increasing harmony in the patient's life.

The energy medicine modality known as Polarity Therapy (PT) uses the three principles of vata, pitta and kapha along with the five elements of ether, air, fire, water, and earth to detect elemental imbalances in a client's body and biofield. Fire imbalances feel warm and inflamed. Water imbalances feel damp and sluggish. Earth imbalances feel dense and immobile, whereas air imbalances are difficult to calm and soothe. Each element has a polarity, exhibiting a positive pole, a negative pole and a neutral pole (See table 1).

Pole	Air	Fire	Water	Earth
Positive(+)	Shoulders	Eyes/Brow	Breast/Chest	Neck
Neutral(φ)	Kidneys	Solar Plexus	Ovaries/Testes	Bowels
Negative(-)	Ankles	Thighs	Feet	Knees

Table 1. The four elements and associated poles and body parts.

Schematics of humans in the fetal position show the connection between different areas of our body and how these connections form triangles of positive, negative and neutral poles for each element (See figures 1-4).

Earth represents the solid state of matter. It expresses stability, permanence, and rigidity. Physical manifestations of earth are bones, teeth, cells, and tissues. Its associated sense is smell. Earth is considered a stable substance. In the fetal triangle earth is represented by the positive pole (neck) and astrological sign Taurus [April 22 – May 21]; neutral pole (bowels) Virgo [August 22 – September 21]; and negative pole (knees) Capricorn [December 22 – January 21] (figure 1).

Figure 1

If the earth element becomes contracted we experience sleepiness, becoming tired shortly after rising. Contracted earth can also cause constipation, knee pain, lethargy, and easy weight gain. Stubbornness, greed, and materialism are all expressed in our biofield as contracted earth. When earth is contracted we can feel neck and low back pain, anxiety and fear, get stuck in old patterns, and grow tumors and fibroids. When earth is expanded it causes inappropriate boundary issues, such as inappropriate touch, inappropriate words, and invasion of another's space. Expanded earth causes paranoia and worry and also contributes to the development of diabetes.

Chronic and incurable diseases of the earth element include osteoarthritis of the knees, lower back, chronic neck and cervical vertebrae problems, as well as colon, blood and bone cancers. If you are an earth sign Taurus (neck), Virgo (bowels) or Capricorn (knees) chronic disease in your associated body part can lead to a debilitating quality of life.

Water represents the liquid state and characterizes change. Water is necessary for the survival of all living things. Between 60 – 70 percent of the human body is made up of water. Water is a substance without stability. Water in our blood, lymph, along with other fluids move between our cells and through our vessels, bringing energy, carry away wastes, regulating temperature, bringing disease-fighting cells, and carrying hormonal information from one area to another. Water's associated sense is taste. It is represented by the positive pole (breasts/chest) and astrological sign Cancer [June 22 – July 21]; neutral pole (generatives - ovaries for women, testes for men), astrological sign Scorpio [October 22 – November 21]; and negative pole (feet) Pisces [February 22 – March 21] (see figure 2).

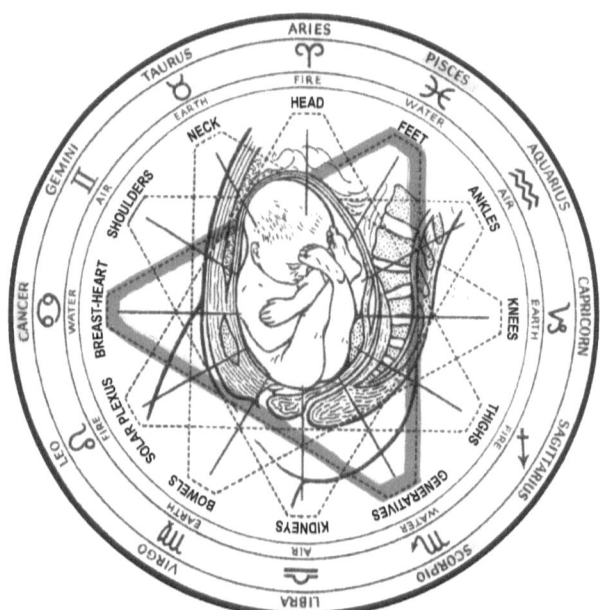

Figure 2

Contracted water is responsible for addictions, possessiveness and attachment. It can also cause the development of excess mucus, endometriosis, swelling, PMS symptoms, prostate trouble, hip and foot pain, intense menstrual cramps, and lymphedema. Contracted water element causes renal failure (due to electrolyte imbalances), swollen brain cells, hypotonic dehydration (due to sodium deficiency), and high concentrations of sodium. Expanded water causes edema, swollen feet and ankles, immune deficiency, compulsive behavior, compulsive crying, living in the past, emotional imbalances, feelings of being overwhelmed and weight gain. Chronic and incurable water diseases include bladder and kidney disorders, chronic edema, lymphoma, and breast, ovary and testicular cancer. If you are a water sign Cancer (breast/chest), Scorpio (ovaries/testes), or Pisces (feet), a chronic disease in your associated body part can lead to a debilitating quality of life.

Fire is the power to transform solids into liquids, to gas, and back again. It possesses the power to transform the state of any substance. Within our body, fire energy binds the atoms together. It converts food to fat (stored energy) and muscle. Fire transforms food into energy. It creates the impulses of nervous reactions, our feelings, and even our thought processes. Fire is considered a form without substance. Its associated sense is sight. It is represented in the body by the positive pole (eyes and brow) and astrological sign Aires [March 22 – April 21]; the neutral pole (solar plexus) Leo [July 22 – August 21]; and negative pole (thighs) Sagittarius [November 22 – December 21], (see figure 3).

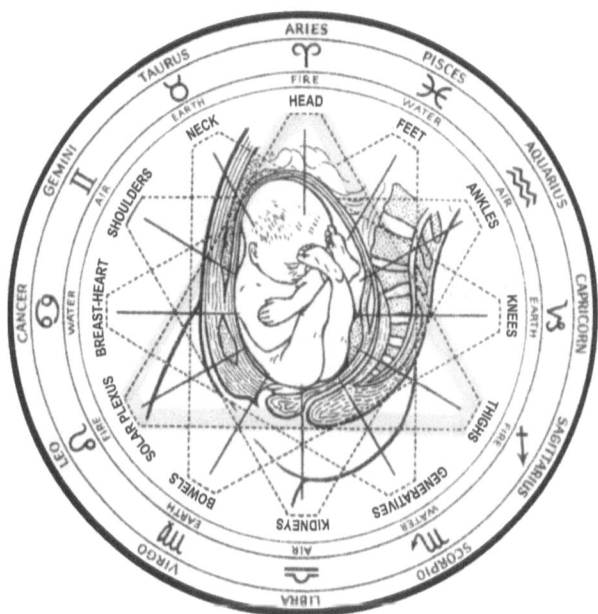

Figure 3

Fire imbalances include emotional instability ranging from cowardice to rage management problems. Contracted fire can cause acid reflux, gallbladder and liver problems, inflammation, boils and infections, eye problems and headaches, lack of clarity and seething emotions. Expanded fire causes ego/power trips, blaming others, explosive rage, loud voice, eye redness and indigestion problems. Chronic and incurable fire diseases include migraine headache, blindness and eye disorders, chronic indigestion, as well as liver, stomach and pancreatic cancer. If you are a fire sign Aries (eyes/brow), or Leo (solar plexus – liver, stomach, gall bladder, and pancreas), or Sagittarius (thighs), chronic disease or health problems associated with your body part can lead to a debilitating quality of life.

Air is the gaseous form of matter which is mobile and dynamic. Within the body, air (oxygen) is the basis for all energy transfer reactions. It is the key element required for fire to burn. Air is existence without form. Its associated sense is touch. It is represented by the positive pole (shoulders) and the astrological sign Gemini [May 22 – June 21]; neutral pole (kidneys) and sign Libra [September 22 - October 21]; and negative pole (ankles) Aquarius [January 22 – February 21](see figure 4).

Figure 4

Air imbalances include jittery uncontrolled movement in the body and in thought patterns. Contracted air presents shoulder pain, shallow breathing, skin problems, feelings of jealousy, lack of compassion and judgment. Expanded air produces dry itchy skin, weak ankles, impatience, scattered and rapid talking. Chronic and incurable air diseases include Chronic Obstructive Pulmonary Disease (COPD), emphysema, kidney disease, chronic shoulder or ankle pain, neurological disease, lung and esophageal cancer. If you are an air sign Gemini (shoulders), Libra (kidneys) or Aquarius (ankles), chronic disease or problems in your associated body part can lead to a debilitating quality of life.

Ether is the space through which all energy moves. It is the source of all matter and the space in which it exists. Ether is also the distance that separates matter. The chief characteristic of ether is sound. Sound represents the entire spectrum of vibration. Ether is represented by all the astrological signs. Contracted ether presents joint problems and brain tumors, whereas expanded ether presents as mental and emotional instability, dementia, and Alzheimer's disease.

There is a cycle of interaction in all five of the elements, in bodily functions, in the environment, and in our unconscious and in our spiritual body. To understand how these five elements interact, it is important to understand the five senses of sound, touch, color, taste and smell and how they interact. Sound, through vibration, produces the Ether element. Ether, which figures prominently in Ayurveda, is considered to be a cosmic substance that is in all people and all things. The movement of touch combined with Ether produces air. The energy of visible color (sight) combined with the energies of sound and touch produce fire. The energy of taste combined with the essence of sound, touch and color produces water. Finally the energy of smell incorporates the qualities of sound, touch, color, taste and smell. These elements enter the body through ingested foods and fluids which continue the state of transformation. Only death ends the cycle. In the body these elements form the humors or doshas. Each dosha has 3 stages: expansion, contraction and equilibrium or neutrality. Balancing the elements of life is the first step towards healing the body. The fastest way to shift the energy in our biofield is to shift our thought patterns. The second fastest way to shift energy in our biofield is to change the food (fuel) we eat. Healthy thoughts and healthy food are the basis for a healthy biofield.

References

1. Chopra, A., Doiphode VV. A., *Ayurvedic medicine—core concept, therapeutic principles, and current relevance.* Medical Clinics of North America., 2002. **86**(1): p. 75-88.
2. Stone, R., *Polarity Therapy.* Summertown, TN., 1986.

CHAPTER TEN
Before Conception to Birth

In order to understand how or why we develop disease, it is important to understand who we are – how we are made. Disease does not manifest from thin air. Disease is cultivated and nurtured to reach destructive levels. Without understanding the environment in which disease thrives, we have no way of detecting or preventing its manifestation. Disease has an energy fingerprint - certain frequencies. It leaves its mark in our biofield at every level. For us to understand disease, we must first understand ourselves.

The energetics of physiology

There is an ongoing argument that life either begins at conception or life begins at the time of birth. Some say our soul first enters through our embryonic stem cells, others argue that a soul would not enter an incomplete body it only enters a body that is autonomous from its mother. Many argue there is life after death because we are eternal therefore we live forever, having a beginning, but no end to our existence. It is my position that we exist long before we are conceived. Not only do we continue living after our physical body dies we existed long before our body was a thought. We incarnate for a purpose – call it education, call it exploration - we are here to live out a mission, which is to leave the earth better off than when we got here.

Our soul is made of energy, our minds are made of energy, our emotions are based on different energetic frequencies; the cells of our body are made of energy. Conception occurs in a dynamic state. It is an idea as well as an act of procreation. Medicine has known for some time how conception leads to birth, with the understanding that when sperm unites with an egg, it provides stimulus for the entire growth process. A single cell begins to undergo a process of self-replication and transforms into an organism of undifferentiated cells. It is well understood that cells begin to differentiate by reading the genetic code of DNA through transcription (via RNA) – differentiating this information in the development of the growing human embryo taking the shape of nerve, bone, muscle, fat, cartilage, connective tissue, etc., then migrating to appropriate positions to form a complete human body. What DNA does not explain however, is how newly differentiated cells know where to appropriately position themselves in a baby's body. How does a stem cell know whether to become a toenail or an eyelash? Embryonic stem cells seem to know what tissue to morph into by following some invisible code. This information system is known as morphogenesis or morphogenetic biofield instruction. A morphogenetic field is an organism's spatial organization of growth from embryogenesis through adulthood and is guided by a holographic energy template known as the human biofield or simply as *The Field*[1].

How does the Field work?

Quantum physics and experiments in high-energy particle physics have shown that all matter is energy[2]. Quantum mechanics challenges the view that separation implies physical independence. Just because two objects are not close together does not mean they do not affect one another. Quantum mechanics transcends space and even long range quantum connections can function despite spatial separation. According to the laws of quantum physics two objects can be far apart in space but it is as if they are a single entity. In physics this is known as *entanglement*. Entanglement implies a holographic order to the human organism[3] and *Superstring Theory* explains how quantum mechanics works in a field. The mechanics between the physical body and higher energy systems is the unique part of our subtle anatomy known as *The Field*[1]. This subtle matter is composed of both particles and waves that connect our observable physical body with our invisible higher body. The spatial organization of our cells is ordered by a complex mapping system known as our morphogenetic field or biofield which organizes the development of our physical body. According to Richard Gerber, MD this field is a holographic energy template that carries the coded information for the spatial organization of the fetus as well as a roadmap for cellular repair in the event of damage to the developing organism[4]. The field consists of lines of energy or force fields each with its own wavelength creating a template to form the entire human body.

Each type of tissue-to-be is represented, differing from other types, because the energy, of which it is an end-product, is itself on another frequency. Thus the body structure, muscular and vascular tissues; the nerves, the brain, and other substances are all represented in the etheric mold by currents of energy on specific frequencies. The emitted vibrations of the surrounding matter cause atoms to enter into differing molecular combinations to produce various types of tissue. These molecules are attracted toward lines of force, settling into their appropriated places in the growing body by virtue of sympathetic vibration or mutual resonance.

To better understand the human organism, we must acknowledge a fundamental physics principle that energy of different frequencies can coexist within the same space without destructive interference. This same concept applies to physical, mental, emotional and spiritual energies which can co-exist in the human body without creating interference the same way varied frequencies such as radio, television, cell phones and satellite frequencies can all exist within the same space without interfering with each other. When multiple signals are sent over a single transmission channel, the process that keeps the signals from interfering with one another is called multiplexing. In frequency domain multiplexing, each signal is given a unique carrier frequency. These frequencies must be assigned so that adjacent signals do not overlap. They are separated by a frequency interval equal at least to the signal bandwidth. Following this fundamental law is an organized pathway of information systems that dictate how the energy of the human body takes shape and form. Cells, tissues, organs, all body parts align according to their designated frequency.

From Source to Form (macrocosm to microcosm)

In ancient Egyptian, Traditional Chinese Medicine (TCM) and Eastern Ayurvedic traditions life is thought of as the movement of conscious energy. From the Universal Life Force come subtle fields that step down into denser physical form. This vital force energy is saturated with conscious awareness arising from Source or God which has been understood for thousands of years in the Hindu philosophy of Ayurveda. It is also understood in the Chinese tradition known as the Tao or "The Way". Source expressing as 'Tao' is known as the neutral essence of all life in TCM and as Sattvas in Ayurvedic medicine.

Before conception, energy steps down from higher vibrations to denser forms following a holographic model of pulsing energy[5].

The Movement of Opposites (yin and yang)

Once intent for birth has been established energy steps down or differentiates by lowering its frequency. This is possible due to a pushing and pulling away from its central core (see figure 1).

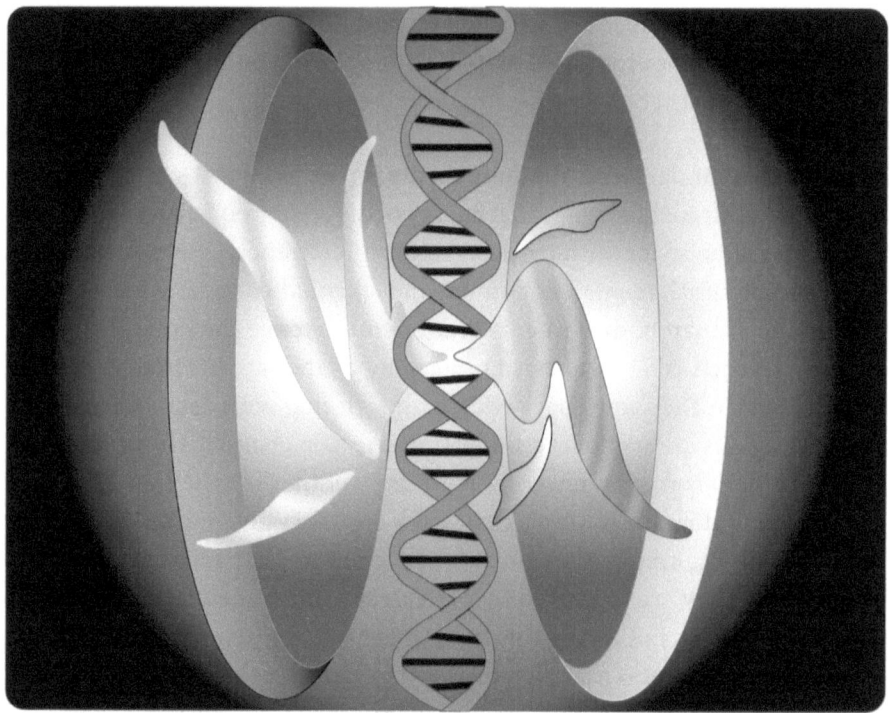

Figure 1. Movement of opposites known as yin and yang.

Movement of opposites is implicit in all creation (night/day, sun/moon, heat/cold, wet/dry, male/female, etc), causing polarity, which is expressed as yin and yang[6]. The ancient medical text of internal medicine *Nei Ching* states that "the entire universe is an oscillation of the forces of yin and yang"[7]. Yang is viewed as the male principle – active, generative, associated with sun, light and the creative principle of life. Yin is viewed as the female – passive, receptive, associated with the moon, the dark and intuition. The separation of the yin and yang from Source is known in other cultures as Ida (yin), Pingala (yang) and Shushumna (Source); Rajas (yang), Tamas (yin) and

Sattvas (Source); Father (Source), Son (yang), Holy Ghost (yin); Id (yang), Ego (yin) and Superego (Source). The dualistic principle of yin/yang extends into all aspects of life cycles and processes as oscillations between polar opposites. Both are necessary to reach a balanced steady state – a dynamic equilibrium within a universe of constant change. Chinese philosophy views a healthy life as one which contains an even balance of the forces of yin and yang. Maintaining equilibrium between these polarities is said to initiate perfect health in mind, body and Spirit. Polar equilibrium applies to our individual organs as well. An imbalance in the flow of meridian energy leads to subsequent organ and blood pathology[8].

The Spiritual [or Causal] Body

The next step in the process of evolution from Source to form is referred to as our "Higher Self"; also known as the Causal or Spiritual body, it is composed of subtle substances of the highest vibrational frequency of an individual entity. The Causal Body initiates abstract ideas and concepts. Causal consciousness is the essence of a subject while the mental aspect studies the subject's details. Consciousness is a form of energy that continually evolves to higher levels of complexity and understanding[5].

Subtle energy in the Causal Body is known to the Chinese as "chi" and to Hindus as "prana". This energy is a manifestation of the life-force which animates and energizes all living systems. Causal representation of an ongoing inner process of thought is concerned with the "meaning of life", as concrete thoughts and ideas are manifested on the physical plane. According to Richard Gerber, MD, chi or prana is negatively entropic in nature[4]. It moves through an organism toward states of increased order and greater cellular energy balance. When the flow of life energy to a particular system or organ is deficient or unbalanced, patterns of cellular disruption occur. Gerber explains:

> Entropy is a term which describes the tendency toward disorder of a system – the greater the entropy, the greater the disorder. Most systems tend toward increasing positive entropy and more disorder over time (i.e. things tend to fall apart). Living systems display the property of negative entropy, or a tendency toward decreasing disorder of the system. They take in substances which are broken down to elements which are less organized then build them up

into systems which are more organized. Living organisms take in raw material and energy, and self-organize them into complex structural and physiological subcomponents. Here life force (vital energy) of the system seems to be associated with negative entropic characteristics. When the body dies and the life force vacates the physical form, the remaining unoccupied shell returns to microorganisms, to its raw constituents in characteristic positive entropic fashion. The etheric body, a self-organized holographic energy template, also demonstrates negative entropic properties[4].

To view life as negatively entropic defies human logic in as much as we see apples ripen and fall off trees to rot. We never see rotten apples become edible and return to the tree. But contrary to this system, our causal body is self-sustainable because it is in constant perpetual motion. Although our body will eventually grow old and die, it was designed to regenerate itself at every level. Our soul however is not designed to die, but to exist eternally. This vital force field is the energy through which the human body regenerates itself during the lifetime of the individual. According to Dr. Randolph Stone, founder of Polarity Therapy, "there is a dimension of human physiology which is the domain of Spirit. The spiritual dimension is the energetic basis of all life, because it is the spiritual dimension which animates the physical framework. The unseen connection between the physical body and the subtle forces of Spirit holds the key to understanding the inner relationship between matter and energy"[9].

The Mental Body

Stepping down from the Spiritual body, energy begins to form the Mental Body, which is the mechanism through which the self manifests and expresses concrete intellect. Consciousness itself is a form of energy. It is the highest form of energy integrally involved in the life process[10]. Mind, consciousness, intellect and thought create experiences that all play a part in the mental body. Franklin Sills, author of *The Polarity Principle*, explains "for energy of the mental realm to have its effects upon the physical, a cascade effect occurs. The mental energies affect matter in the astral body which are then transmitted to the etheric body and finally to the physical body"[5].

In theoretical physics, superstring theory explains the important role electrons, quarks, and other particles play in our biofield, how these particles are not individualized points of matter but instead waves of energy working in synchronistic formation. Every particle is composed of a tiny filament of energy shaped like a little string or thread, hundreds of trillions of times smaller than a single atomic nucleus. Just as violin strings vibrate, these super strings vibrate, each producing a different musical tone or frequency[3]. Different particle properties or string-like characteristics create the fabric of The Field. According to biophysicist James Oschman, "the body as a whole, the organs, tissues, cells, organelles, including the nucleus and the strands of genetic material, DNA, can be viewed as a continuous and unbroken fabric: a matrix within a matrix within a matrix". He refers to this continuum as the connective tissue/cytoskeleton/nuclear matrix, or simply *The Living Matrix*[11].

Formation of the Chakras and the Nadi Points

Links between our subtle energy anatomy and our physical nervous system are known as the subtle threads of the Nadi system. Nadi points are interlaced with physical nerves in the body, relaying magnetic currents from the chakras at various energetic levels[12]. The chakras use higher frequency energy to form human structure. Chakras "step-down" energy the same way electrical transformers "step-down" electrical voltage from high power lines to usable current in houses. Chakras decrease the flow of high frequency energy into subtle energy channels of the cell structure in the physical body. Chakras are connected to each other and to various areas in the physical body through energetic threads known as nadis. Together with the chakras, the nadis become conduits to the nerves, veins, vessels or arteries composing the subtle body of the human biofield. Nadis process vibrational energy of specific frequencies, while our chakras translate the effect of etheric, astral and higher vibrational frequencies into the glands of our endocrine system[9].

The nadis are different than the meridians, which have a physical counterpart in the meridian duct system. The idea of nadis first appeared in ancient Indian Ayurvedic texts known as the *Upanishads* (7th – 8th century c.e.). The *Kshurika-Upanshiad* and later the *Hathayogapadikpa* mention as many as 72,000 nadi channels in the subtle anatomy of the human body. The most important nadis are the Shushumna, Ida and Pingala[13]. The nadis of one energy body are connected to the nadis of neighboring energy bodies through the chakras. Nadi channels are interwoven within the physical nervous system. Nadis affect the nature and quality of nerve transmission within the brain, spinal cord and peripheral nerves. Dysfunction at the level of the chakras and nadis can be associated with pathological changes in the nervous system. Like Chinese meridians, the nadis constitute channels of vital energy known as prana (chi).

The chakras are able to affect our emotions and behavior through hormonal influences in brain activity. Emotional stability is a function of properly working chakras and integrated subtle bodies. Chakras regulate the flow of vital energy into different organs of our body. When functioning properly, they establish strength and balance in our physical body. Abnormal chakra functioning can create weakness in every area of the body[14].

There are seven major chakras (any many minor chakras) associated with the physical body. Anatomically, each major chakra is associated with a particular endocrine gland and with a major nerve plexus in the spinal column (See Table 1).

Chakra Number	Name	Location	Related Element	Related Color	Related Endocrine Gland	Related Nerve Plexus/Ganglia
1st	Root	Base of spine	Earth	Red	Adrenals	Coccygeal
2nd	Sacral	Sacrum	Water	Orange	Ovaries/Testes	Pelvic/Sacral
3rd	Solar Plexus	Below sternum and above navel	Fire	Yellow	Pancreas	Solar/Lumbar
4th	Heart	Above the heart	Air	Green	Thymus	Cardiac/Pulmonary Thoracic
5th	Throat	Throat	Ether	Blue	Thyroid	Cervical
6th	Third Eye	Center of forehead	-	Violet	Pituitary Gland	Sympathetic/Superior cervical
7th	Crown	Top of head fontanel	-	Magenta	Pineal Gland	Parasympathetic

Table 1
Associations of the Seven Chakra Centers

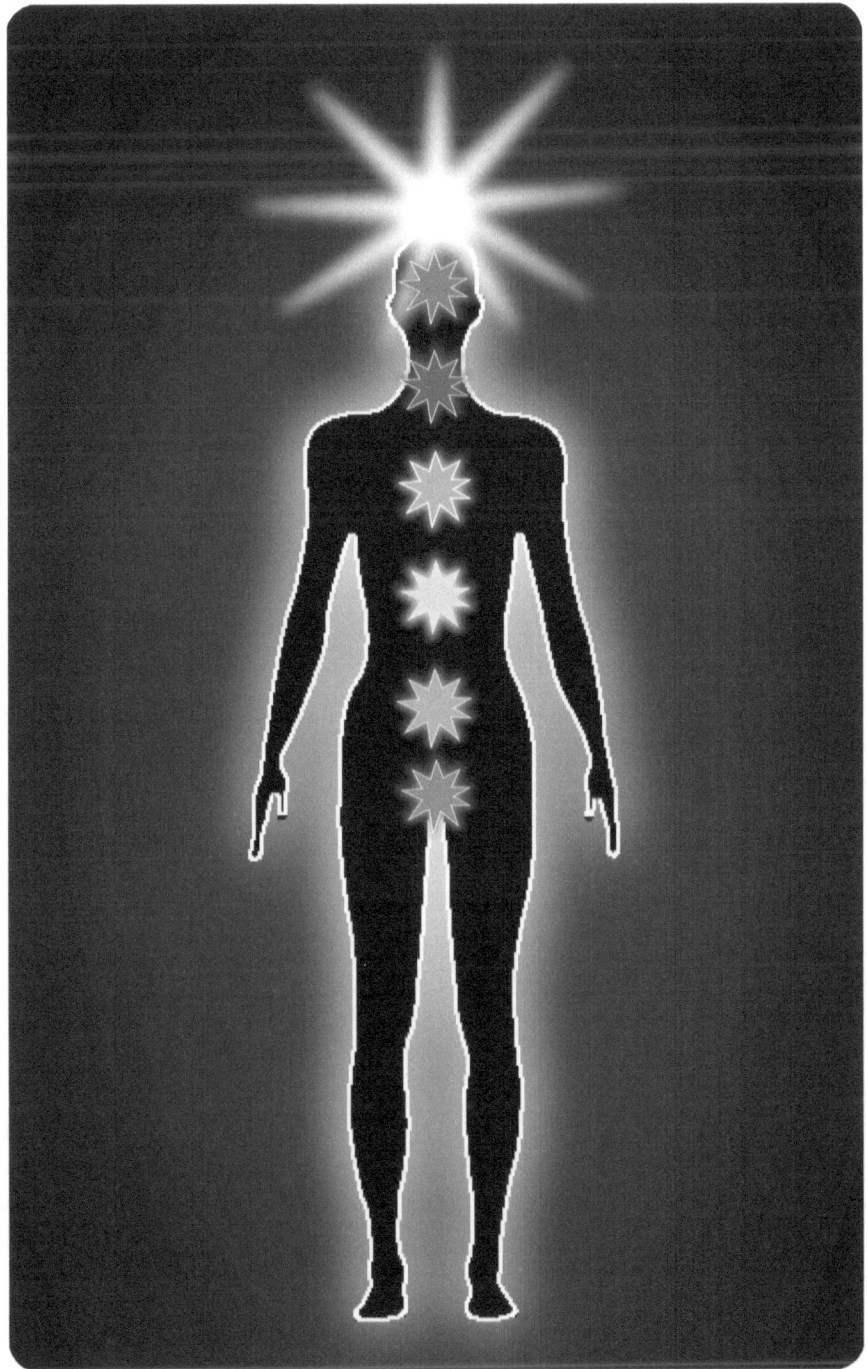

Figure 2. *The major chakras are situated in a vertical line ascending from the base of the spine to the head.*

The seven chakras include: the first chakra, also known as the root chakra; the second chakra, also known as the sacral chakra; the third chakra, also known as the solar plexus chakra; the fourth chakra, also known as the heart chakra; the fifth chakra, also known as the throat chakra; the sixth chakra, also known as the third eye chakra; and the seven chakra, also known as the crown chakra. Our first chakra is red in color and is associated with the adrenal glands in the endocrine system. It connects us with the earth and associates us with our family or clan. Our root chakra is represented by the earth element. The second chakra, also known as our sacral chakra, is orange in color and associated with the reproductive organs (ovaries/testes). It is our source of reproduction and creativity. Our sacral chakra is represented by the water element. The third chakra is yellow in color and is associated with the pancreas, and is the body's power center. It keeps our mind focused and motivated. It is represented by the fire element. The fourth chakra is green in color and is associated with the heart chakra. It is the center for love of self and for others. It is represented by the element of air. The fifth chakra is blue in color and is associated with the thyroid. It gives us a voice to speak. It is represented by the element of ether. The sixth chakra is associated with the pituitary gland and is purple in color. It provides psychic vision and connects us with the mental plane or 5th dimension. The seventh chakra is associated with the pineal gland and is magenta in color. I have seen it sparkle in a diamond faceted-like rainbow of magenta, silver and gold. It connects us with God, the universe and the spirit world.

The chakras are all associated with specific endocrine glands which control the genetic expression of our physical body through the functioning of the central nervous system. Endocrine glands secrete hormones directly into our blood stream providing information to our body through a signaling system of different mechanisms. Hormonal effects are slow to initiate, lasting from hours to weeks in contrast to our nervous system which reacts very quickly and its effects are short lived. The fastest nerve signals travel at speeds that exceed 100 meters per second and between 8 – 13 cycles per second. Harmonic frequencies of the chakras however can reach the range of many thousands of cycles per second[15]. The hormonal activity of the major endocrine glands is dependent upon the influence of their associated chakra and its energetic potential.

The Formation of Oval Fields

At each chakra fields of energy arise which are referred to as *oval fields*. These fields originate from inside and permeate around the chakras. As the founder of Polarity Therapy, Dr. Randolph Stone explains "for energy to move there must be a field to support its movement"[9]. Without a field to move through, energy cannot flow. These pulsating 'motor' fields provide a medium through which other energy patterns move and evolve. The chakras can be thought of as a sensory feedback system. If more energy is required in a particular pattern, that information is relayed to the chakras through the oval fields where an appropriate adjustment is made.

Marma Points

From the center of this organization of energy patterns comes the formation of marma points. Marmas are various subtle energy compositions relative to the tissues they create. They are defined as bone, tendon, muscle, nerve, or vein, relative to the channels that carry the doshas (the biological elements of vata, pitta and kapha) along channels that carry our thoughts and emotions. While many marma points are on the surface of the body, others are internal like the heart and liver, which are large marma regions. Many marma points are located on the peripheral regions of the body (i.e. arms, legs, hands and feet). The head has the greatest concentration of marma points, with special marmas governing the eyes, ears, nostrils, mouth and brain. Many marma points can also be found along the front and back of the torso as well[16].

The Astral Body

Our astral body is a subtle body made up of energy / matter of a higher frequency than the subtle body of etheric matter. It too is superimposed on the physical-etheric framework, but unlike the etheric body, which supports and energizes our physical body, our astral body functions as a vehicle of consciousness, which can exist separately, yet connected to our physical body[4]. Our astral body is involved in the expression and repression of our emotions and often referred to as the emotional body. Astral centers are receivers and transmitters of astral energy which is stepped down and passed on to the chakras, where through the nadis energy becomes transduced into nerve and glandular functions. Since our astral body is involved in emotional expression, our astral chakras provide a subtle energy connection through which our emotional state can either disrupt or enhance our health depending on the frequency. Glandular and hormonal function occurs at the level of cellular activity and hormones play an integral part in the emotional expression of our personality representing our feelings[17]. Emotional imbalances are due to neurochemical disturbances in the brain as well as to abnormal patterns of energy flow within our astral body and our chakras. The degree to which we are affected by our emotions governs the nature of our personality. Dysfunction in the astral body caused by emotional imbalances can impair the flow of energy through the chakras, eventually resulting in glandular problems and physical illness[4].

Our Etheric Body

Our etheric body assists us in the movement of the Life-Force, prana or chi through our DNA. The difference between the physical body and the etheric body is in their characteristic frequencies. These same characteristics also distinguish the etheric body from the astral, mental and spiritual bodies. Our etheric body is not completely separated from our physical body. It too uses nondestructive energy patterns (known as frequency domain multiplexing) which provide specific frequencies to transfer information from one body to another.

Research suggests that our etheric body forms a type of holographic magnetic grid which communicates with electrically based cell matter of the physical body through the meridian system. Separate from the central nervous system, researchers speculate this mechanism introduces a separate channel of cell communication with calcium waves playing the role of the second messenger[18]. The mechanical wave, the acoustic shear wave, and the calcium wave turn out to have come from the same source, and these different forms of waves appear ideally suited to describe what the ancient Chinese called chi or qi, the life-force energy[19].

Richard Gerber, MD describes the etheric body as

> "a holographic energy template that guides the growth and development of the physical body. It maintains order and determines structural patterning within the cellular matrix of the physical body. The etheric energies provide a wave guide upon which to organize cell structure and function. Distortions of coherent energy patterns of the etheric template can lead to abnormal cell growth. Diseases appear in the etheric field weeks and even months prior to manifesting in the physical body"[4].

When energy medicine practitioners become aware of these incoherent patterns in the human biofield they can stabilize them before they manifest as errors in cell communication[20]. It is very difficult for some people to release incoherent patterns that are fixed in their biofield. Because our biofield is organized by our astral body / emotions, people who are attached to their emotions embed them into their soul. Our emotions should be as fluid as the events we experience in our lives. Viewing every situation as a learning experience by embracing the good with the bad will allow for the ultimate healing environment. "Soul emotions and the energy that accompanies them can stick around literally for lifetimes because they are related to our survival", says Valerie Hunt. "They form ways of interacting with physical reality in order to preserve the body itself". Disease manifests in the physical body only after energy disturbances become crystallized in the subtle structural patterns of the higher frequency bodies[15]. Through the etheric body energy frequencies regulate cell growth patterns in the physical body. The etheric body carries spatial information determining how the fetus is to develop in utero and also the structural data for growth and repair of the adult organism in the event of damage or disease.

Energy Meridians

The meridian system interacts with the nervous system through a series of energy transduction steps which regulate the higher energy systems by influencing cells and tissues in our physical body. The meridian system is the first physical link established between our etheric body and our developing physical body. The meridians are electrical circuits connecting exterior acupuncture points to deeper organ structures[21]. The meridians supply the nutritive chi (prana) to bodily organs. Meridian changes reflect dysfunction occurring at an etheric level.

In the developing fetus, sufficient energy in meridian circuits is essential to keeping their energy balanced with respect to one another. There is a harmonic flow in the chi as it passes through the twelve meridians supplying energy to the internal organs. While scratching a small area of skin certain people often simultaneously experience an additional sensation at a remote site on the body. For instance, while scratching your nose you may feel a similar sensation in your lower back. This is known as mitempfindung. The transmission of mitempfindung along acupuncture meridians may involve a series of C-fiber-Merkel cell relays, with the final referred itch sensation caused by substance P release triggering mast cell degranulation. This sensation follows the meridian lines in the body[21].

Researchers discuss the importance of acupuncture practice utilizing diagnosis and distribution of various meridians and connecting channels based on meridian theory. The meridian system is considered the basic anatomy for acupuncture, so corresponding pathways and related symptoms of different channels play a key role in differentiation. This is known as meridian-related pattern differentiation. Research is showing how acupuncture using meridians can be applied clinically in cases such as whiplash injury, intervertebral disc herniation, oculomotor nerve paralysis, and eczema[22]. Meridians form early during embryogenesis in order to act as spatial guides for the growth and development of newly forming blood and lymph systems in the body. As blood vessels develop they grow around the meridians. Embryological studies show meridian ducts are formed within fifteen hours of conception[23]. This meridian system exerts an influence on the migration and spatial orientation of the internal organs. Because the meridians connect to each cell's DNA, the meridian system plays an important role in both the replication and differentiation of all cells in the body[24].

Our Physical Body

Our physical body is the densest component of the many interactive energy fields that make up the human organism. Each of these fields or higher frequency bodies is connected to the physical cellular structure through a complex network of energy threads[4]. If the energy in the etheric field becomes distorted, physical disease soon follows. Many illnesses initiated in the etheric body later manifest in the physical body as organ pathology. Because it is composed of both energy and matter, our physical body has both particle and wave-like properties. These electromagnetic properties of our physical body can be influenced by electromagnetic field stimulation.

The Energetics of DNA

Cellular biology demonstrates that every cell contains a copy of the master DNA blueprint with enough information to make an entire human body from scratch. The holographic model helps us to understand structures from the level of the single cell all the way up to levels of cosmic order[25]. Through holography we understand the hidden qualities of matter at both the microscopic and macroscopic levels. *As above – so below,* also known as the *Principle of Correspondence* is visible at the microscopic level where cells of living organisms display organizing principles demonstrating that every piece contains the whole. At the macroscopic level the growth of the entire organism is guided by an invisible etheric template similar to a hologram in three dimensions[9]. There is well documented evidence suggesting DNA receives and transmits information in the heart's electrical rhythms and in the oscillation of the DNA molecule itself[26].

Subatomic Particles

All matter is composed of innumerable types of particles. In the developing human fundamental particles of matter known as subatomic particles exist within atoms, electrons, neutrons, and protons, which are made up of smaller particles known as quarks[3]. Subatomic particles are similar to electrons in that their subcomponents resonate with different vibrational frequencies. These different vibrational states have many octaves and harmonics which are referred to as healthy states or diseased states depending on their frequencies[15].

Atoms

At the atomic level, resonance occurs via electrons vibrating around the nucleus in energetically defined orbits. In order to move an electron from a lower orbit to a higher orbit, a quantum or packet with a determined amount of energy is required. An atomic orbital is a mathematical function that depends on the coordinates of the electron. It describes the wave-like nature of an electron in an energy state which is referred to as a wave function. An electron will only accept energy of the appropriate frequency to move from one energy level to another because "like energy attracts like". If the electron falls from a higher orbit to a lower orbit, it will radiate energy of that very same frequency. This required atomic frequency is referred to as the "resonant frequency". The vibrational mode of the physical body is a reflection of the dominant frequency at which it resonates[4]. "Although the energy level of humans varies from moment to moment, day to day, the physical body tends to vibrate at a particular frequency. Because of its vibratory properties, the physical body can be affected by electromagnetic forces. A key principle of resonance and MRIs (magnetic resonance imaging) is that hydrogen atoms are being stimulated by the transfer of energy of a specific frequency. Because of this principle of 'resonance specificity' – *like energy attracts like* - atoms possess unique chemical and physical properties"[4].

Molecules

Molecules are combinations of atoms. As the smallest units of a compound, molecules are affected by electromagnetic forces and are defined by their unique energetic frequencies[27]. Molecules can be influenced by energy medicine therapies due to the molecular arrangement of our physical body, which is a complex network of interwoven energy fields. Biofields associate the physical and chemical structures of the human body through cell communication or cell signaling. Cell signaling, which is an exchange of protein molecules between cells, explains how molecules of emotion can affect health and well-being. The work of neuroscientist and pharmacologist Candice Pert emphasizes emotions are not only derived through a feedback of the body's environmental information, but also through self-consciousness. The mind uses the brain to generate "molecules of emotion" to obtain feedback from the organism's environmental information[28]. Stressful environments promote sympathetic nervous system (fight or flight) activation, while peaceful environments encourage the regeneration of the organism through the parasympathetic nervous system (rest and digest). Serenity promotes health and wellness while stress causes illness and disease. The difference occurs in the organism's (person's) ability to deal with both internal as well as external stress.

Organelles

An organelle is a specialized subunit within a cell having a specific function and usually separately enclosed within its own lipid membrane (see figure 3).

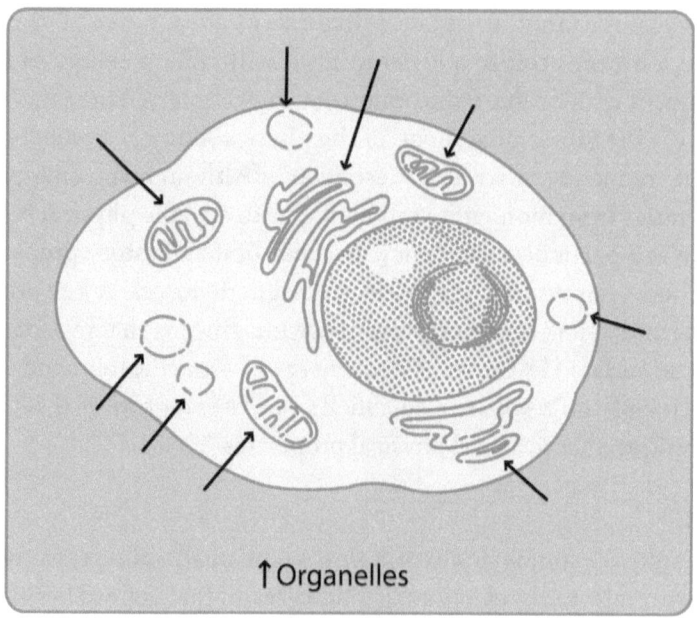

↑ Organelles

Figure 3

The name organelle comes from the concept that these structures are to cells what an organ is to the body. In addition to the nucleus, there are many organelles inside of a cell, which carry out cell functions. One important cellular organelle is the ribosome. Ribosomes participate in protein synthesis. The transcription phase of protein synthesis takes places in the cell nucleus. After this step is complete, the messenger RNA (mRNA) leaves the nucleus and travels to the cell's ribosomes, where translation occurs, (meaning the way in which a protein sequence can be encoded by a nucleic acid sequence, given by the genetic code). Another important cellular organelle is the mitochondrion. Mitochondria are often referred to as the power plants of the cell because many of the reactions that produce energy take place in the mitochondria[29]. In addition to supplying cell energy, mitochondria are involved in a number of other processes, such as cell signaling, cell differentiation, cell death, as well as the control of the cell cycle and cell growth. Mitochondrial membrane functions can be characterized by small-amplitude oscillations which correlate behavior over a wide range of frequencies[30]. Abnormal frequencies are a sign of cellular energy imbalances within our physical body[24]. Also important in the life of a cell are the lysosomes. Lysosomes are organelles containing enzymes which aid in the digestion of nutrient molecules and other materials. These organized parts of a cell have unique chemical functions which actively seek environments that support their survival while simultaneously avoiding toxic or hostile environments[31]. This type of signal communication is fundamental to the survival of the organism.

Cells

The cell is the basic structural and functional unit of all known living organisms. All cells have recognizable features such as a plasma membrane which protects a cell from its outside environment. The cell membrane regulates the movement of water, nutrients and wastes into and out of the cell. At the center of the cell is the nucleus (see figure 4).

It contains cell DNA which is the genetic code that coordinates protein synthesis. Although the cells of our body have enzymatic control systems for self-maintenance and replication, they are guided by natural energy patterns of a higher frequency. Normal human body frequencies vary depending on the tissue. Neuroscientists have found the normal frequency range of the human brain oscillates between a few Hertz (Hz) and ~20 Hz[32]. Cell membrane potentials governing biological transport systems like CL^-, Na^+, and K^+ ions pass through membrane channels that have voltage potentials. Most cell membrane potentials average around -70 electron volts (eV). The membrane potential of neurons averages around -90 eV. Average heart beat frequencies at ~120 beats per minute (bpm) would equate to around 2 Hz and human visual image processing averages around 60 Hz in well lit conditions. If measured, every organ in our body would produce a certain oscillatory frequency when healthy and a different frequency when we are sick. This variance from our normal healthy frequency renders our immune system unable to properly defend our body against bacteria, viruses and toxic chemicals. If however our body gets treatment for the frequency it needs to defend itself against disease, then the cells resonate at normal healthy frequencies. When our immune system is resonating at healthy frequencies, it can throw off illness allowing our body to return to a healthy level of homeostasis[24].

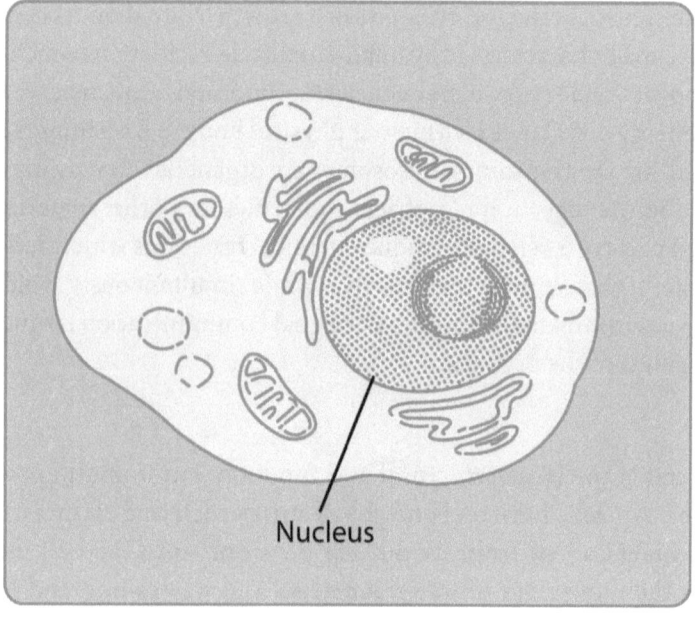

Figure 4

Tissues

Biological tissue is a group of cells that carry out a specific function. The development, repair and regeneration of living systems in our body come from cell tissue. Organs are formed by the functional grouping together of multiple tissues. Animal tissues can be grouped into four basic types: muscle, nerve, epithelial, and connective. Muscle tissue functions to produce force and cause motion, either locomotion or movement within internal organs.

Muscle tissue is separated into three categories: visceral or smooth muscle, found in the inner linings of organs; skeletal muscle, which is attached to bone providing gross movement; and cardiac muscle found in the heart, allowing it to contract and pump blood throughout the organism. Nerve cell or neural tissue makes up the central nervous and peripheral nervous systems. In the central nervous system, neural tissue forms the brain and spinal cord; and in the peripheral nervous system forms the cranial nerves and spinal nerves[33]. This includes the motor neurons. The epithelial tissues are formed by cells that cover organs such as the surface of the skin, the air passages, the reproductive systems, and the inner lining of our digestive tract. The cells comprising an epithelial layer are linked through a semi-permeable barrier between the external environment and the organ it covers. Epithelial tissue protects organs from microbes and pathogens, injury, and fluid loss. Connective tissue is made up of closely packed collagen fibers separated by non-living material, called extracellular matrix. Cells create their own type of matrix depending on the necessary function of the cell. Connective tissue gives shape to organs and also holds them in place. It supports and binds tissues together. Unlike epithelial tissue, connective tissue typically has cells scattered throughout the extracellular matrix.

Because tissue is made of cells with membrane potential, tissue is capable of being influenced by electromagnetic fields. Tissue forms in either normal healthy patterns or distorted diseased patterns based on the cell signaling involved in creating the tissue. Electromagnetic fields can directly influence the electrophysiology of tissue by perturbing cell membrane potential[34, 35]. Researchers have detected shifts in membrane potential along the length of cells in response to an electrical pulse. For elongated tissue cells, or groups of tissue cells that are coupled electronically by gap junctions, significant hyper-polarizations and de-polarizations can result from millisecond applications of electric fields in the 10-100 mV/cm range[36].

Organs

Organs are groups of tissues that perform a certain function in the body. Clustered within each organ are nerve cells called ganglia. The ganglia exchange information with the brain and coordinate data with the other systems. Each organ within our body has its own energy frequency. Organs of similar frequency tend to be clustered together in the same body regions or are linked together in a special physiologic system. There are separate meridians for each major organ in the body[37]. There is a heart meridian, a liver, meridian, a spleen meridian, a bladder meridian, a small intestine meridian, a kidney meridian, a gall bladder meridian, a lung meridian, the circulation-sex meridian, a large intestine meridian, and a stomach meridian. If the energetic potential or chi within an organ is not stable it will be unable to complete the natural meridian circuit. This can affect the adjacent organ in the meridian system. For instance congestive heart failure can lead to cardiac cirrhosis of the liver which can lead to venous congestion and enlargement of the spleen. These systems are all connected by the liver meridian whose endpoint is the heart chakra[4].

Organ Systems

In the human body an organ system is a group of organs that work together to perform a certain task or function in our body. The major physiological organ systems are the **cardiovascular** system which consists of blood, blood vessels and the heart; the **lymphatic** system, which consists of lymph nodes and vessels, the thymus, and the spleen; the **digestive** system, which breaks down food polymers into smaller molecules to provide energy to the body. This system consists of the mouth, teeth, tongue, stomach, liver, pancreas, intestines, and rectum. The **endocrine** system which maintains growth in the body consists of the pituitary gland, pineal gland, hypothalamus, ovaries, testes, and thyroid gland. The **integumentary (skin)** system protects the internal structures of the body from damage. It prevents dehydration, stores fat and produces vitamins and hormones. Also included in the organ systems are the **skeletal** system, the **circulatory** system, the **respiratory** system, the **muscular** system, the **reproductive** system and the **nervous** system. These systems are all guided energetically by the meridians of the human biofield. Meridians not only feed vital energy to their related organs, they also reflect any pathological disturbance in these organs, providing health care practitioners with a convenient and highly accurate tool for diagnosis[38].

Nervous System

Our biofield meridians follow the pathway between our nervous and circulatory systems in order to fuel our body with vital energy known as chi. This energy extends directly into the molecular system of our body. Nerves are the dense physical manifestation of energy threads underlying neurons, pointing to the subtle energy links in our physical body[4]. *There is a strong connection between our astral body and the energy of our nervous* system. Our reaction to our environment is well established in our parasympathetic,

sympathetic and central nervous systems. Reaction to stressors in our environment either cause illness and disease or promote health and wellness depending on the experience[28].

In this book, *The Body Electric*, Robert O. Becker, MD discusses the DC current system of glial cells involved in self healing electrical feedback loops that influence the production and transmission of action potentials of these nerves[39]. The glial cell network functions as an interconnection between the meridians and the nervous system. Glial cells are non-neuronal cells that maintain homeostasis, form myelin, and provide support and protection for neurons in the brain, and for neurons in other parts of the nervous system such as the autonomic nervous system[40]. Nerve cells are constantly releasing neurotransmitters into the synaptic gaps between themselves and the neurons they contact. The DC potential carried over these cells energetically affects the nerves they surround by influencing the pre-synaptic sites. The cell membrane's electrical potential determines the responsiveness of each neuron in releasing neurotransmitters on cue. These signals can be influenced and manipulated by exogenous fields[41].

Central Nervous System

Our central nervous system (CNS) is the part of our nervous system that integrates information it receives from all parts of our body. It represents the majority of our nervous system and consists of the brain and the spinal cord. All cells of the nervous system are made up of neurons. Nerve processes consist of axons and dendrites which are able to conduct and transmit signals. Axons typically carry signals away from the cell. They are long nerve processes that branch out to convey signals to various parts of our body. Dendrites typically carry signals toward the cell. Axons and dendrites are bundled together into what are called nerves. Nerves send signals between the brain, spinal cord, and other body organs through nerve impulses. Neurons are classified as either motor, sensory, or interneurons. Motor neurons carry information from the central nervous system to organs, glands, and muscles. Sensory neurons send information to the central nervous system from internal organs or from external stimuli. Interneurons relay signals between motor and sensory neurons.

Diseases of the central nervous system include encephalitis and poliomyelitis, neurodegenerative diseases such as Alzheimer's disease and amyotrophic lateral sclerosis (ALS or Lou Gehrig's disease), autoimmune and inflammatory diseases such as multiple sclerosis or acute disseminated encephalomyelitis, and genetic disorders such as Krabbe's disease and Huntington's disease (Huntington's chorea). Cancer of the central nervous system can cause severe illness and, when malignant, can have very high mortality rates. Neuromuscular diseases such as stroke, Parkinson's disease, multiple sclerosis, muscular dystrophy, myasthenia gravis and Creutzfeldt-Jakob disease are all neurological diseases.

Astral energies have their impact on our brain and nervous system through their connection with the etheric body and its integration with our physical body. When disease occurs in the nervous system our personality can become trapped in a non-expressive body (i.e. stroke, paralyzation, etc.)[4]. Structural, biochemical or electrical signaling errors leading to or from the brain, spinal cord, or nerves, can result in paralysis, muscle weakness, poor coordination, loss of sensation, seizures, confusion, pain and altered levels of consciousness. "Dissociation of the astral form from the physical body is a manifestation of some type of primitive reflex which protects our consciousness from traumatic experience. There is a constant state of resonance along our spinal column between the medulla oblongata and the coccyx [craniosacral connection]; properties of the pineal gland resonate between these points"[4].

Sympathetic Nervous System

The sympathetic nervous system stimulates our fight or flight mechanism. The sympathetic nervous system (SNS) is one of the three parts of the autonomic nervous system, along with the enteric (which directly controls the gastrointestinal system) and parasympathetic nervous system (PNS). The SNS mobilizes our body's stress mechanisms by stimulating the fight-or-flight response. Sympathetic nerves originate at the base of our neck and are responsible for stabilizing mechanisms in our body. Fibers from the SNS innervate tissues in almost every organ system, providing at least some regulatory function to stimuli such as pupil diameter, gut motility, and urinary output. The SNS response acts primarily on the cardiovascular system transmitting impulses indirectly through catecholamines secreted from the adrenal medulla. When threatened by our environment, our pupils dilate so we can see better, our vessels dilate to send more oxygen to our muscles, and we rush adrenaline and cortisol throughout our body to give us more energy[4]. This occurs whether the threat is real or imagined.

SNS disruption can come from either environmental stimuli or internal thought patterns. SNS dysfunction can cause various illnesses, including genetic disorders[42]; congenital disorders[43]; infections[44]; lifestyle or environmental health problems including malnutrition[45]; and brain, spinal or nerve injury[46]. The problem may start in another body system that interacts with the nervous system. For example cerebrovascular disorders involve brain injury due to problems with the blood vessels supplying blood to the brain, and autoimmune disorders that involve damage caused by our body's own immune system. Exposure to this type of stress can also age you prematurely. A research study showed where children exposed to continuous violence experience wear and tear on their DNA similar to what is seen in aging. These children were physically abused by an adult or bullied frequently. They also had heightened risk of mental and physical disorders as adults[47].

Parasympathetic Nervous System

The parasympathetic nervous system (PNS) is one of the three main divisions of the autonomic nervous system (ANS). The ANS is responsible for the regulation of internal organs and glands which occurs subconsciously. The parasympathetic system is responsible for the stimulation of "rest-and-digest" activities that occur when our body is calm. These activities include tears, salivation, sexual arousal, digestion, urination and defecation. For the same reason constant stress displayed in the sympathetic nervous system can be detrimental to our health and well-being, the serenity displayed in our parasympathetic nervous system regenerates vital cells and organs of our body, keeping our immune system strong and able to fight off disease.

Parasympathetic and sympathetic divisions of the nervous system function in opposition to each other. In a healthy body, the PNS naturally decreases the activity of the SNS, bringing our body back to a healthy, peaceful equilibrium. This natural opposition is better understood as complementary rather than antagonistic. The SNS typically functions during events that require quick responses. The PNS reaction typically does not require immediate reaction.

Complex Multi-Cellular Organisms

The complex multi-cellular organism constitutes our physical body. At this stage of life, Source has completed its journey to form, manifesting the human body en toto. Where once there was pure energy is now a mature fetus. An imbalance of yin and/or yang appears in our meridian circuitry as abnormalities of organ pathology. Disease states are improved by correcting these energetic imbalances in our body's meridian circuitry through the incoherent patterns that precede cellular dysfunction and disorganization[37]. Each system works in harmony with the other systems along a hierarchical axis of energy flow. The uninhibited flow of chi through our body's energy channels (chakras, nadis, meridians and nervous system) maintains the vitality of our physical body. Because our spiritual body and our mental body both feed energy into our astral/emotional body (which funnels down into our etheric/physical body), healing from the spiritual and mental levels produce longer lasting results than healing from either the astral (psychological) or etheric (physiological) levels.

References

1. McTaggart, L., *The Field: the quest for the secret force of the universe*. Harper Collins Publishers, 2002. **New York**.
2. Lomax, A., *Charged particle therapy: the physics of interaction*. Cancer J, 2009. **15**(4): p. 285-91.
3. Greene, B., *The Elegant Universe*. New York, 2003. **W.W. Norton & Company**.
4. Gerber, R., *Vibrational Medicine*. Bear & Company, Santa Fe, NM, 1998.
5. Sills, F., *The Polarity Principle: energy as a healing art*. North Atlantic Books, 2002. **Berkley, CA**.
6. Stone, R., *Polarity Therapy*. 1986. **2**(CRCS Publications): p. Summertown, TN.
7. Ching, N., *The Su Wen of the Huangdi Neijing (Inner Classic of the Yellow Emperor*. Yellow Emperor's Inner Canon, Used for over 2000 years(Ancient Chinese medical text - fundamental doctrinal source for Chinese medicine).
8. Jia, J., Yu, Y, Deng, J, Robinson, N, Bovey, M, Cui, Y, Liu, H, Ding W, Wu, H, Wang, X, *A review of Omics research in acupuncture: the relevance and future prospects for understanding the nature of meridians and acupoints*. J Ethnopharmacol., 2012. **140**(3): p. 594-603.
9. Stone, R., *Polarity Therapy*. Summertown, TN., 1986. **2**.
10. Tiller, W., *What Are Subtle Energies?* J Sci Explor, 1993. **7**(3): p. 293-304.
11. Oschman, J., *A biophysical basis for acupuncture*. Society for Acupuncture Research, 1995. **Rockville, MD**.
12. Judith, A., *Wheels of Life: a user's guide to the chakra system*. Llewellyn Publications, 2002. **St. Paul, MN**.
13. Rao, A., *'Mind' in Indian philosophy*. Indian J Psychiatry, 2002. **44**(4): p. 315-325.
14. Shang, C., *Emerging paradigms in mind-body medicine*. J Altern Complement Med., 2001. **7**(1): p. 83-91.
15. Hunt, V., *Infinite Mind: Science of the Human Vibrations of Consciousness*. Malibu Publishing Company, 1996. **Malibu, CA**.
16. Zarrilli, P., *Three bodies of practice in a traditional South Indian martial art*. Soc Sci Med., 1989. **28**(12): p. 1289-309.
17. Pert, C., Snyder, S, *Opiate Receptor: Demonstration in Nervous Tissue*. Science, 1973. **179**(4077): p. 1011-1014
18. Chang, K., Wong, T, Wong, T, Leung, A, Chung, J, *Effect of acupressure in treating urodynamic stress incontinence: a randomized controlled trial*. Am J Chin Med. , 2011. **39**(6): p. 1139-1159.

19. Yang, E., et al., *Ancient Chinese medicine and mechanistic evidence of acupuncture physiology.* Pflugers Arch - Eur J Physiol, 2011. **Aug 26**: p. Epub ahead of print.

20. Hunt, V., *The promise of bioenergy fields: an end to all disease.* http://www.spiritofmaat. com/archive/nov1/vh.htm, 2012 retrieved May 15.

21. Silberstein, M., *Do acupuncture meridians exist? Correlation with referred itch (mitempfindung) stimulus and referral points.* Acupunct Med., 2012. **30**(1): p. 17-20.

22. Chen, Y., et al., *Effective acupuncture practice through diagnosis based on distribution of meridian pathways & related syndromes.* Acupunct Electrother Res., 2011. **36**((1-2)): p. 1-18.

23. Lee, T., *Thalamic neuron theory: meridians=DNA. The genetic and embryological basis of traditional Chinese medicine including acupuncture.* Med Hypotheses, 2002. **59**(5): p. 504-21.

24. Fraser, P. and H. Massey, *Decoding the Human Body-Field: the new science of information as medicine.* Rochester, VT, 2008. **Health Arts Press**.

25. Talbot, M., *The Holographic Universe.* Harper Collins Publishers, 1991. **New York**.

26. Rein, G. and R. McCraty, *Modulation of DNA by coherent heart frequencies.* Proceedings of the 3rd annual conference of the International Society for the Study of Subtle Energies and Energy Medicine, 1993. **Monterey, CA**(June).

27. Hwang, M., Ni, X, Waldman, M, Ewig, C, Hagler A, *Derivation of class II force fields. VI. Carbohydrate compounds and anomeric effects.* Biopolymers, 1998. **45**(6): p. 435-68.

28. Pert, C., *The Molecules of Emotion:The Science Behind Mind-Body Medicine.* Simon & Schuster, Inc., 1997. **New York**.

29. McBride, H., Neuspiel, M, Wasiak, S, *Mitochondria: more than just a powerhouse.* Curr. Biol, 2006. **16**(14): p. R551–60.

30. Aon, M., Cortassa, S, O'Rourke, B, *Mitochondrial oscillations in physiology and pathophysiology.* Adv Exp Med Biol, 2008. **641**: p. 98-117.

31. Kerfeld, C., Sawaya, M, Tanaka, S, Nguyen, Cv, Phillips, M, Beeby, M, Yeates, T, *Protein structures forming the shell of primitive bacterial organelles.* Science 2005. **309**(5736): p. 936–938.

32. Baker, S., *Oscillatory interactions between sensorimotor cortex and the periphery.* Curr Opin Neurobiol, 2008. **17**(6): p. 649-55.

33. Alberts, B., Bray, D, Lewis, J, Raff, M, Roberts, K, Watson, J, *Molecular Biology of the Cell.* Garland Publishing, Inc, 1983. **New York**.

34. Irmak, M., *Multifunctional Merkel Cells: their roles in electromagnetic reception, fingerprint formation, Reiki, epigenetic inheritance and hair form.* Medical Hypothesis, 2010. **75**: p. 162-168.

35. Escoffre, J., Dean, D, Hubert, M, Rols, M, Favard, C, *Membrane perturbation by an external electric field: a mechanism to permit molecular uptake.* Eur Biophys J, 2007. **36**(8): p. 973-83.

36. Cooper, M., *Membrane potential perturbations induced in tissue cells by pulsed electric fields.* Bioelectromagnetics, 1995. **16**(4): p. 255-62.

37. Eden, D., *Energy Medicine.* Penguin Putnam, Inc., 1998. **New York**.

38. Reid, D., *Guarding the Three Treasures.* Shambhala Publications, Inc., 1994. **Boston, MA**.

39. Becker, R., *The Body Electric: Electromagnetism and the Foundation of Life*. New York, 1985. **William Morrow & Co., Inc.**

40. Jessen, K., Mirsky, R, *Glial cells in the enteric nervous system contain glial fibrillary acidic protein*. Nature, 1980. **286**: p. 736 - 737.

41. Pennisi, G., Ferri, R, Lanza, G, Cantone, M, Pennisi, M, Puglisi, V, Malaguarnera, G, Bella, R, *Transcranial magnetic stimulation in Alzheimer's disease: a neurophysiological marker of cortical hyperexcitability*. J Neural Transm, 2011. **118**(4): p. 587-98.

42. Currie, G., Freel, E, Perry, C, Dominiczak, A, *Disorders of blood pressure regulation-role of catecholamine biosynthesis, release, and metabolism*. Curr Hypertens Rep, 2012. **14**(1): p. 38-45.

43. Furness, J., Poole, D, *Nonruminant Nutrition Symposium: Involvement of gut neural and endocrine systems in pathological disorders of the digestive tract*. J Anim Sci, 2012. **90**(4): p. 1203-12.

44. Garcia, A., Fels, R, Mosher, L, Kenney M, *Bacillus anthracis lethal toxin alters regulation of visceral sympathetic nerve discharge*. J Appl Physiol, 2012. **112**(6): p. 1033-40.

45. Landsberg, L., *Feast or famine: the sympathetic nervous system response to nutrient intake*. Cell Mol Neurobiol, 2006. **26**(4-6): p. 497-508.

46. Fargali, S., Sadahiro, M, Jiang, C, Frick, A, Indall, T, Coglian,i V, Welagen, J, Lin, W, Salton, S, *Role of Neurotrophins in the Development and Function of Neural Circuits That Regulate Energy Homeostasis*. J Mol Neurosci, 2012, Nov;48(3):654-9.

47. Shalev, I., Moffitt, T, Sugden, K, Williams, B, Houts, R, Danese, A, Mill, J, Arseneault, L, Caspi, A, *Exposure to Violence During Childhood is Associated with Telomere Erosion from 5 to 10 Years of Age: A Longitudinal Study*. Molecular Psychiatry, 2012. **April 24th. doi:10.1038/mp.2012.32**.

CHAPTER ELEVEN
Cell Communication and Disease

Cell communication plays a pivotal role in the origination and prevention of disease. Cells communicate by signaling one another. They signal by sending small protein molecules in the form of chemical messengers and electrical impulses that are transmitted by specific cells and received by their target cells (see figure 1). Cell signaling is responsible for the communication between our cells and tissues that regulate our body's systems and processes. Cell signals can be carried from cell to cell, from cell to organ or from our brain to other parts of our body through our bloodstream. The role of cell signaling is vital to our overall health and well being.

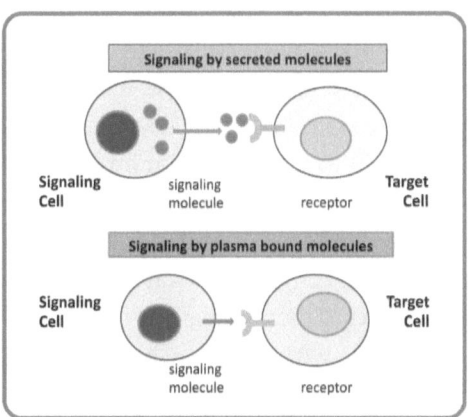

figure 1. Cells communicate by signaling one another, Top: intercellular signals by small protein molecules in the form of chemical messengers and electrical impulses; and bottom: extracellular signaling which is response to a ligand that binds to a specific receptor.

Communication between cells is required to coordinate diverse cell activity. Cells have two main compartments: 1) the nucleus which contains the genes, and 2) the cytoplasm which contains the cell substructure known as the organelles. Research has shown that the coordination of cells and their signals profoundly regulates the transfer of genetic information from the nucleus of the cell to the cytoplasm and through the membrane of the cell into our body[1]. Understanding how this mechanism works is the key to controlling the growth of cancer cells, as well as analyzing genetic diseases such as cystic fibrosis and Huntington's disease.

There are about 210 distinct human cell types and between 50 and 75 trillion cells in the human body. They are capable of interacting, communicating, and performing complex tasks in unison[2]. This capability enables our body to adapt to its environment which is the key to our survival. Cells respond to environmental stimuli by moving toward a stimulus if it is life enhancing; moving away from a stimulus if it is life threatening; or ignoring a stimulus if it is benign. When we perceive threatening stimuli, we engage in a protection response by shutting down and moving away from the threat. If we are constantly in protection mode, we cannot grow life enhancing cells in our body. This lack of cell proliferation can lead directly to the death of our organs, systems and ultimately our body[3]. Because our cells constantly send out and receive signals, if a cell fails to signal at the proper time or if that signal fails to reach its target, a disease can result. If a target cell does not respond to a signal, or a cell responds even though it has not received a signal, this can result in disease as well. Most diseases involve at least one breakdown in cell communication[4].

Miscommunication between cells creates many different problems. For example cells in the pancreas release signals called insulin. Insulin signals the liver, muscle and fat cells to store the sugar for later use. In type I diabetes the pancreatic cells (β cells) that produce insulin are lost so the insulin signal gets lost. As a result, sugar accumulates to toxic levels in the blood. Without treatment, diabetes can lead to kidney failure, blindness and heart disease[5]. Type I and type II diabetes have very similar symptoms, but they have different causes. People with type I diabetes are unable to produce the insulin signal. People with type II diabetes can produce insulin but lose the ability to respond to the insulin signal, so it gets ignored. In either case the end result is the same - blood sugar levels become dangerously high.

Multiple sclerosis is a disease in which the myelin sheath around nerve cells in the brain and spinal cord are destroyed. The affected nerve cells can no longer transmit signals from one area of the brain to another. The nerve damage caused by multiple sclerosis leads to muscle weakness, blurred or double vision, difficulty with balance, uncontrolled movements, and depression[6].

A stroke occurs when a blockage forms in a blood vessel, cutting off blood flow to part of the brain. The immediate result is the death of nearby brain cells. But the most damage comes later when the dying cells release large amounts of a signaling molecule called glutamate. Glutamate spreads through the brain and kills cells that were not affected by the blockage. This leads to widespread brain damage. In this case too much cell signaling causes the damage[7].

Despite constant checks and balances, cell communication can break down resulting in uncontrolled cell growth often leading to cancer. Cancer always requires multiple breakdowns in signaling[8]. Many times cancer begins when a cell gains the ability to grow and divide even in the absence of a signal. Under healthy conditions this increased cell growth triggers a signal for self-destruction, but when the cell also loses the ability to respond to signals that kill the cell, it divides out of control, forming a tumor. Later cell communication events cause blood vessels to grow into the tumor, enabling it to grow larger. Additional signals allow the cancer to spread to other parts of the body.

Errors in cell communication are responsible for disease.

Cell communication operates at both local and long-distance levels. Signaling molecules are emitted in the form of gases, fatty acid derivatives, or proteins. An example of a gas that operates on a locally based signaling system is nitric oxide (NO). NO is a signal for lowering blood pressure. Hormones are an example of long-distance signaling molecules that must travel through the circulatory system to reach their target cells. These types of cell signals include testosterone, estrogen, progesterone (endocrine signals) and insulin. Cells also communicate through short-range signals which are secreted by one cell and cover a short distance before they are recognized and interpreted by another cell. An example of short-range cell signaling is neurotransmitters which travel across the tiny spaces between adjacent neurons or between neurons and muscle cells. This form of cell communication occurs through electrical and chemical signaling. The points at which electrical signals are converted into chemical signals and then back into electrical signals are called synapses.

Errors in cell communication also cause inflammatory responses to

117

repeat uncontrollably causing chronic inflammation in tissue. Chronic inflammation and errors in cellular information processing are responsible for inflammatory diseases[9] and diseases such as cancer[10], autoimmune diseases[11], diabetes[12], myocardial infarction[13], and Alzheimer's disease[14]. Inflammation also plays a role in heart disease because the immune system attacks high-density lipoproteins (HDLs) which protect against the development of atherosclerotic coronary heart disease[15]. Chronic inflammation eventually damages the arteries, which can cause them to burst. In fact, inflammation is so closely associated with heart disease that many doctors now use a test for inflammation called CRP (C-reactive protein) to assess a person's risk of heart attack. Research shows that CRP can predict the risk of heart attack and stroke as well or better than cholesterol levels[16]. Inflammation is also a key characteristic of pain and edema[17]. Studies report certain inflammatory cell signals are involved in the initiation and persistence of pathologic pain through activated nociceptive sensory neurons[18, 19].

While studying a certain inflammatory pathway that triggers inflammation I came to believe that chronic inflammation is the cause of all disease. When our body becomes inflamed through either trauma or by pathogens, it attacks the invading microbes with neutrophils, macrophages, B- and T-cells, to ingest the invaders, create antibodies, clean up and regenerate fresh cells where previous ones were destroyed.

When there are errors in cell communication the inflammatory response can turn on us. Because cell behavior is dependent on the interaction of proteins and their complementary signals, either the proteins are defective or the signals are distorted when an error occurs[20]. Since approximately 5 percent of the world's population is born with birth defects, meaning they have mutated genes that code for dysfunctional proteins[21], then 95 percent of disease can be attributed to a dysfunction in cell signaling. Errors in cell communication can be attributed to trauma (physical, mental or emotional); toxins that distort signals between our nervous system and targeted cells; or our body's inability to accurately perceive environmental information and engage in life-sustaining behavior[20]. There are many inflammatory pathways and many opportunities for miscommunication between cells.

The process by which information moves inside the cytoplasm of a cell is also a form of cell signaling. Signaling occurs in response to a ligand that binds to a specific receptor on the cell surface causing changes inside the cell. During an inflammatory response certain lymphocytes known as monocytes leave the blood and mature into macrophages that enter the tissue around the injured or infected area. Macrophages can either stimulate inflammation or suppress it by releasing chemical signals that alter the behavior of other cells[17]. They phagocytose or ingest the offending pathogen and clean up debris creating an environment for new cells to regenerate into healthy tissue. An acute inflammatory response is important to our immune system because it regulates the cells that attack microbes such as bacteria, viruses and toxic chemicals. The ability of cells to perceive and correctly respond to their environment is the basis of development, tissue repair, and immunity as well as normal tissue homeostasis. In addition to fighting pathogens, the immune system monitors the health of cells and disposes of cells that have been injured and killed[22]. When microbes such as bacteria or viruses invade or breach epithelia and enter our tissues or blood stream, they are attacked by specialized lymphocytes called phagocytes, and several plasma proteins[22, 23]. The phagocytes known as macrophages give off cell signals called cytokines to warn of the invasion. Cytokines are inflammatory proteins that react to tissue injury and infection and are synthesized in a wide range of biological actions in various tissues[24-26]. They are cell signaling messengers that have specific effects on the interactions between cells, on communication between cells, or on the behavior of cells. It is the pro-inflammatory cytokine cell signaling mechanism that is responsible for the initiation of the inflammatory response.

My doctoral dissertation focused on the effect of a pulsed electromagnetic field (PEMF) on a pathway consisting of the pro-inflammatory cytokine tumor necrosis factor (TNF), also known as TNF-alpha (TNF-α); a nuclear factor kappa B (NF-kB), a nuclear factor kappa-light-chain-enhancer of activated B cells and protein complex that controls the transcription of DNA; and the gene A20 (TNFAIP3). Energy therapies such as PEMF can change the regulation of cell signaling processes of proinflammatory markers such as Tumor Necrosis Factor alpha (TNF-a) and Nuclear Factor kappa B (NF-kB) which initiate inflammation[27]. They are able to do this because cell signaling is frequency dependent. Signals moving from one cell to another behave in certain ways depending on the frequency they receive. The sub extra low frequency electromagnetic field (sub ELF-EMF) in the 0.3-30 Hertz (Hz) range were the most effective in regulating the inflammatory response on the TNF-NFkB-A20 pathway. A20 is a gene whose expression is rapidly induced by the pro-inflammatory cytokine tumor necrosis factor (TNF). A20 inhibits NF-kB activation as well as TNF-mediated apoptosis, or cell death[28]. TNF along with IL-1 activates a major signal transduction pathway called the Jak-

STAT signaling pathway. This is an important signaling pathway involving the Jak's which are kinases, and also STAT's which are transcription factors (STAT stands for "signal transducer and activator of transcription") which turn on a cascade of gene regulation in response to cytokine stimulation. In molecular biology and genetics, a transcription factor is a protein complex that binds to specific DNA sequences, controlling the flow (or transcription) of genetic information from DNA to mRNA[29]. A gene is a hereditary unit consisting of a sequence of DNA that occupies a specific location on a chromosome and is the blueprint for a particular characteristic.

My dissertation involved inducing macrophages with an endotoxic substance that comes from *E coli* bacteria known as lipopolysaccharide or LPS. When the macrophages were induced with LPS they became inflamed, swelling up and aggregating to amplify their cell signaling ability. Results of our research showed that PEMF down regulates proinflammatory cell signaling in inflamed tissue and creates an environment where cells heal and regenerate themselves. Whether the inflammatory response is initiated from pathogens or tissue trauma, the exogenous cell environment plays a significant role in the ability of our cells to regain normal function and heal tissue. Electromagnetic field (EMF) affects cell signaling in ways that are much faster and more effective than chemicals such as pharmaceuticals. Studies show that at extremely low frequencies EMF does not have the debilitating side effects of pharmaceuticals[30]. If the outcomes of my PEMF study are reproducable by the hands of an energy medicine practitioner, we will soon understand how energy therapies affect cell signaling and ultimately health and wellness.

How Can Energy Therapies affect Cell Signaling?

This question was the basis of my dissertation. While studying inflammatory response and disease, I realized the basis of *all* non-genetic based disease is chronic inflammation. Perhaps even genetic diseases could be triggered by inflammation[31-33]. Since inflammation is signal-mediated, cells respond to signals that are sent immediately after our body has been invaded. In a chronic stage inflammatory response, macrophages continue to tear down and rebuild tissue causing the recurrence of inflammation. Chronic inflammation severely compromises our immune system.

To counteract chronic inflammation energy medicine (EM) can be administered in the human biofield through either a device or the hands of an EM practitioner. An interesting characteristic of energy emission from any living organism is that it stays somewhat organized in its fields. It has a tendency to remain stable and does not randomly dissipate[34]. Cell as well as biofield vibrations are like tuning forks, acting as both transmitters and receivers of vibration coming from their environment. Many research studies have detected the frequency limit of cell oscillations to be only 30 Hz[35-39], which is the same frequency range coming from the hands of EM practitioners[40]. They resonate at specific harmonic pitches when we are healthy. When we are not healthy a non-coherent type noise vibrates from our cells and our biofield. This suggests a subtle resonance involved in the healing process. This naturally occurring phenomenon known as resonance has also been measured in the hands of Therapeutic Touch (TT) practitioners using a superconducting quantum interference device (SQUID)[41]. A SQUID is a magnetometer used to detect very weak biomagnetic fields. A biomagnetic field oscillating between (0-30 Hz) was detected coming from the hands of these practitioners. This frequency range is considered to be a sub extra low-frequency electromagnetic field (sub ELF-EMF)[42]. When I practice Polarity Therapy I feel a vibration coming from my hands and my clients tell me they feel my hands getting very warm as I work. More research is needed, but there is a trend showing these low frequency treatments affect the inflammatory response in vivo as well as in vitro[43]. I believe that all cell signaling is frequency dependent. When the practitioner has aligned her energy field to a harmonic healthy frequency, the client's field will resonate at that same frequency. Energy therapies such as Polarity Therapy (PT), Therapeutic Touch (TT), Healing Touch (HT), Reiki, Qi Gong, Acupuncture, Trager Approach, Bowen Technique, therapeutic massage, chiropractic, as well as pulsed electromagnetic field (PEMF) devices have all produced healing effects (see Chapter *Resources*). TT performed by trained energy healers significantly stimulated the growth of bone, tendon and skin cells in lab dishes[44]. This research showed the hands of a Therapeutic Touch (TT) practitioner not only stopped the growth of bone cancer (osteosarcoma) cells but stimulated the growth of new bone cells called osteoblasts.

Much of the error or miscommunication during cell signaling is caused by the reaction of a cell to its environment[45]. If we are stressed, fatigued, sedentary, immune compromised, eating unhealthy food, or breathing in toxic chemicals we are triggering breakdowns in cell communication. During these unhealthy situations our cells are receiving interfering signals from their environment causing malfunctions in their ability to differentiate, proliferate, and rebuild new tissue. Energy medicine has the capability of reprogramming unhealthy signals into healthy frequencies. While energy therapies can be beneficial in maintaining healthy cell communication, only we are responsible for creating the healthy environment in which our cells thrive.

References

1. Nanduri, J., Tartakoff, A, *The arrest of secretion response in yeast: signaling from the secretory path to the nucleus via Wsc proteins and Pkc1p.* Mol Cell., 2001 **8**(2): p. 281-289.

2. Alberts, B., Bray, D, Lewis, J, Raff, M, Roberts, K, Watson, J, *Molecular Biology of the Cell.* Garland Publishing, Inc, 1983. **New York**.

3. Lipton, B., *The Biology of Belief: Unleashing the power of consciousness, matter and miracles.* Elite Books, 2005. **Santa Rose, CA.**

4. Brooke, M., Nitoiu, D, Kelsell, D, *Cell-cell connectivity: desmosomes and disease.* J Pathol, 2012. **2**(226): p. 158-171.

5. Cubbon, R., Ali, N, Sengupta, A, Kearney, M, *Insulin- and Growth Factor-Resistance Impairs Vascular Regeneration in Diabetes Mellitus.* Curr Vasc Pharmacol, May 2012;10(3):271-84

6. Antel, J., Antel, S, Caramanos, Z, Arnold, D, Kuhlmann, T, *Primary progressive multiple sclerosis: part of the MS disease spectrum or separate disease entity?* Acta Neuropathol, May 2012; 123(5):627-38.

7. Iadecola, C., Anrather, J, *The immunology of stroke: from mechanisms to translation.* Nat Med, 2011 **17**(7): p. 796-808.

8. Ronquist, G., *Prostasomes are mediators of intercellular communication: from basic research to clinical implications.* J Intern Med, Apr 2011, 271(4):400-13.

9. Ohashi, P., *T-cell signalling and autoimmunity: molecular mechanisms of disease.* Nat Rev Immunol., 2002. **2**(6): p. 427-438.

10. Zhang, Q., et al., *Monoclonal antibodies as therapeutic agents in oncology and antibody gene therapy.* Cell Res, 2007. **17**: p. 89-99.

11. Fujimoto, M. and S. Sato, *B-cell signaling and autoimmune diseases: CD19/CD22 loop as a B cell signaling device to regulate the balance of autoimmunity.* Journal of Dermatol Sci, 2007. **46**(1): p. 1-9.

12. Zhang, J., et al., *Insulin disrupts B-adrenergic signalling to protein kinase A in adipocytes.* Nature, 2005. **437**(569-573).

13. Klingenberg, R., Lüscher, T, *Inflammation in Coronary Artery Disease and Acute Myocardial Infarction: Is the stage set for Novel Therapies?* Curr Pharm Des, 2012;18(28):4358-69.

14. Tan, Z., et al., *Inflammatory markers and the risk of Alzheimer disease: The Framingham Study.* Neurology, 2007. **68**: p. 1902-1908.

15. Gordon, T., Castelli, W, Hjortland, M, Kannel, W, Dawber, T, Fielding, C, Fielding, P *High density lipoprotein as a protective factor against coronary heart disease:the Framingham Study.* . Am J Med Genet B Neuropsychiatr Genet., 1977. **62:**. p. 707–714.

16. Madjid, M., Willerson, J, *Inflammatory markers in coronary heart disease.* Br Med Bull, 2011. **100**: p. 23-38.

17. Outtz, H., Wu, J, Wang, X, Kitajewski, J, *Notch-1 Deficiency Results in Decreased Inflammation During Wound Healing and Regulates Vascular Endothelial growth Factor Receptor-1 and Inflammatory Cytokine Expression in Macrophages.* Journal of Immunology, 2010. **185**: p. 4363-4373.

18. Li, W., et al., *Fracture induces keratinocyte proliferation and expression of pro-nociceptive inflammatory mediators.* Pain, 2010. **151**(3): p. 843-852.

19. Akopians, A., Babayan, A, Beffert, U, Herz, J, Basbaum, A, Phelps, P, *Contribution of the Reelin signaling pathways to nociceptive processing.* Eur J Neurosci, 2008. **27**(3): p. 523-37.

20. Lipton, B., Bhaerman, S, *Spontaneous Evolution: our positive future (and a way to get there from here).* Hay House: Carlsbad, CA, 2009: p. 255.

21. Willett, W., *Balancing Life-style and genomics research for disease prevention.* Science, 2002. **296**: p. 695-698.

22. Karavitis, J., Kovacs, E, *Macrophage phagocytosis: effects of environmental pollutants, alcohol, cigarette smoke, and other external factors.* J Leukoc Biol, 2011. **90**(6): p. 1065-1078.

23. Gomperts, B., I. Kramer, and P. Tatham, *Signal Transduction.* Elsevier, 2009(Burlington, MA, Second Edition).

24. Aggarwal, B. and J. Vilcek, *Comparative analysis of the structure function of TNF-alpha and TNF-beta.* Immunol Ser, 1992. **56**: p. 61-78.

25. Akira, S., et al., *Biology of multifunctional cytokines: IL-6 and related molecules (IL-1 and TNF)* FASB J 1990. **4**: p. 2860-2867.

26. Benveniste, E., *Inflammatory cytokines within the central nervous system: sources, function, and mechanism of action.* Amer J of Phys, 1990. **269**: p. C1-C16.

27. Ross, C., Harrison, B, *Regulation of A20 Gene Expression in Response to Pulsed Electromagnetic Field.* in publication, 2012.

28. Opipari, A., Boguski, M, Dixit, V, *The A20 cDNA induced by tumor necrosis factor alpha encodes a novel type of zinc finger protein.* J Biol Chem, 1990. **265**(25): p. 14705–14708.

29. Latchman, D., *Transcription factors: an overview.* Int J Biochem Cell Biol, 1997. **29**(12): p. 1305–1312.

30. Rubik, B., *Bioelectromagnetics and the future of medicine.* Admin Rad J, 1997. **16**(8): p. 38-46.

31. Barnum, C., Tansey, M, *Neuroinflammation and Non-motor Symptoms: The Dark Passenger of Parkinson's Disease?* Curr Neurol Neurosci Rep, 2012. **May 12 ahead of print**.

32. Calvo, A., Moglia, C, Balma, M, Chio, A, *Involvement of immune response in the pathogenesis of amyotrophic lateral sclerosis: a therapeutic opportunity?* CNS Neurol Disord Drug Targets, 2010. **9**(325–30).

33. Khoshnan, A., Patterson, P, *The role of IκB kinase complex in the neurobiology of Huntington's disease.* Neurobiol Dis, 2011. **43**(2): p. 305-11.

34. Dirac, P., *The Principles of Quantum Mechanics.* Clarendon Press, Oxford, 1930.

35. Korenstein, R., Levin, A, *Membrane fluctuations in erythrocytes are linked to mGATP-*

dependent dynamic assembly of the membrane skeleton. Biophysical Journal, 1990. **60**: p. 773-737.

36. Tuvia, S., Almagor, A, Bitler, A, Levin, S. Korenstein R, Yedgar, S, *Cell membrane fluctuation are regulated by medium macroviscosity: evidence for a metabolic driving force.* Proceeding of National Academy of Science USA 94, 1997: p. 5045-5049.

37. Tuvia, S., Moses, A, Nathan, G, Levin, S, Korenstein, R, *β–adrenergic agonists regulate cell membrane fluctuations of human erythrocytes.* Journal of physiology, 1999. **516**(3): p. 781-792.

38. Popescu, G., Badizadegan, K, Dasari, R, Field, M, *Coherence properties of red blood cell membrane motions.* Journal of Biomedical Optics Letters 2006. **11**(4): p. 040503.

39. Popescu, G., Park, Y-K, Dasari, R, Badizadegan, K, Feld, M, *Coherence properties of red blood cell membrane motions.* Physical Review 2007. **E 76**(031902).

40. Zimmerman, J., *Laying on of hands healing and therapeutic touch: a testable theory.* J Bioelectromag Inst, 1990. **2**: p. 8-17.

41. Currents, B., *Hands of TT practitioner measured by SQUID.* Bio-Electro Mag Institute 1992. **2**: p. 8.

42. Seto, A., Kusaka, C, Nakazato, S, Huang, W, Sato, T, Hisamitsu, T, Takeshige, C, *Detection of extraordinary large bio-magnetic field strength from human hand during external QI emmision.* Acupunct Electrother Res, 1992. **17**: p. 75.

43. Fouladbakhsh, J., *Complementary and alternative modalities to relieve osteoarthritis symptoms.* Am J Nurs., 2012. **112**(3 Suppl 1): p. S44-51.

44. Jhaveri, A., Walsh, S., Wang, Y, McCarthy, M, Gronowicz, G, *Thereapeutic Touch Affects DNA Synthesis and Mineralization Of Human Osteoblasts in Culture.* J Orthop Res, 2008. **26**(11): p. 1541-1546.

45. Kuroda, J., Shimura, Y, Yamamoto-Sugitan,i M, Sasaki, N, Taniwaki, M, *Multifaceted Mechanisms for Cell Survival and Drug Targeting in Chronic Myelogenous Leukemia.* Curr Cancer Drug Targets., 2013 Jan 1;13(1):69-79.

CHAPTER TWELVE
Morphogenesis and Holography

Morphogenesis is the biological process that causes the human organism to develop its shape. Morphogenesis is one of the three fundamental aspects of developmental biology along with the control of cell differentiation and cell growth[1]. Certain cell types "sort out", meaning they sort themselves into groups that maximize contact between cells of their same type. The ability of cells to do this comes from differential cell adhesion. Two well-studied types of cells that sort out are epithelial cells and mesenchymal stem cells (MSCs). During embryonic development some cellular differentiation events cause MSCs to become epithelial cells and at other times epithelial cells differentiate into MSCs. Following this epithelial-mesenchymal transition, cells can migrate away from an epithelium and then associate with other similar cells in a new location. This process controls the organized spatial distribution of cells during embryonic development[2].

Our body is formed through morphogenetic processes which use holographic principles to distribute instruction for stem cell differentiation. Holography is the recorded scattering of electromagnetic radiation (EMR) or information from an object later reconstructed so that when an imaging system such as a camera (or human eye) is placed in the reconstructed beam, an image of the object will be seen even when the object is no longer present (see figure 1).

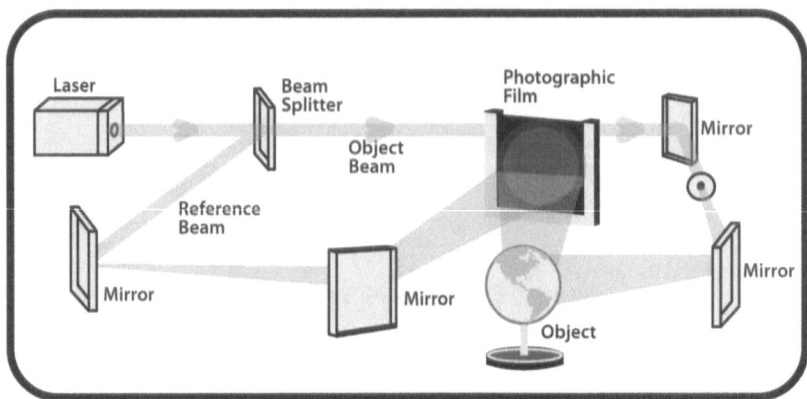

Figure 1. Schematic of a hologram

The image changes as the position and orientation of the viewing system changes in exactly the same way as if the object were still present making the image appear three-dimensional. A holographic recording is not a plain image but consists of random structure of varying intensity and density. Similarly human biofields vary depending on their diverse frequencies. Laser light patterns in holography have a coherent, well organized structure similar to a healthy human biofield[3]. This holographic model of cell morphogenesis allows cells to form the human body by following certain signals to orient themselves.

Morphogenetic Fields

Since the 1920s developmental biologists have suggested that biological organization depends on certain fields called morphogenetic fields. In developmental biology, a morphogenetic field is a group of cells that are able to respond to discrete, localized biochemical signals that guide the human body in the development of specific morphological structures or organs[4, 5]. As an organism develops, embryonic stem cells create blueprints of tissues and organs that are activated within the morphogenetic field[6]. Cells are then assigned to certain groups, for instance cells assigned to the glial field only become brain cells, and cells in a cardiac field will only become heart tissue. The specific programming of individual cells in a field is flexible, for example an individual cell in a cardiac field can be redirected through cell signaling to replace specific damaged or missing cells[7].

Morphogenesis also takes place in mature organisms, in cell culture or inside tumors, due to changes in cell interaction that form tissues. There are several types of molecules which are particularly important during morphogenesis. Morphogens, which are dissipating molecules that spread out and carry signals controlling cell differentiation and bind to specific protein receptors[8], and transcription factor proteins which are an important class of molecules involved in morphogenesis that determine the fate of cells by interacting with DNA. These can be coded for by master regulatory genes that either activate or deactivate the transcription, creating a complementary ribonucleic acid (RNA) copy of other genes. All cells contain physical, chemical and electrical information stored as data[9]. This stored data provides information about a cell's position in our body, initiating individual cell behavior to be synchronized into tissues and organs. Morphogenesis is established during embryonic development. Errors in the morphogenetic process manifest as birth defects. Morphogenetic fields have scalar waves that are organized, structured and contain a great deal of information which is relayed to all cells, tissues, and organs in our body[2]. These field waves predict that each cell knows how every other cell is behaving at any given time. Since we have 50 - 70 trillion cells (depending on our size), all this information could only be processed with the help of structured information fields that have orderliness in space.

Now that molecular biology is one of the most researched fields in biomedicine, a much deeper understanding of human biofields needs to be examined. DNA in the genes of living systems does not carry all the information needed to shape the system, but acts only as a blueprint that fine tunes the morphogenetic fields of previous systems of the same type organism. While evidence supporting the importance of biofields is strong, mechanisms may be explained through information that is not necessarily inside every cell, but outside the cell as well[10, 11]. "What is crucial for morphogenesis and regenerative medicine is the understanding of factors that allow biological systems to self-assemble into consistent, highly complex patterns"[2], says developmental biologist Michael Levin of Tufts University Center for Regenerative and Developmental Biology.

In his article *The wisdom of the body: future techniques and approaches to morphogenetic fields in regenerative medicine, developmental biology and cancer,* Levin suggests that regeneration, development and cancer can all be seen as different aspects of the same cell signaling process[9]. An interesting aspect of this morphogenetic cell signaling process is that the most highly regenerative animals tend to have the lowest incidence of cancer[12, 13]. Regenerative medicine and cancer biology are currently focused on the mechanisms of proliferation control, differentiation and metastasis. Levin also suggests that important advances in regenerative medicine will come with knowing that external signals can activate or modify birth defects or malformed structure after injury if morphogenetic patterns are directly encoded into our morphogenetic field. This allows for manipulation of the structural information - resetting the healthy patterns that are part of the original architectural design[9].

Steady-state ion currents, voltage gradients and electric fields are produced by ion channels and pump proteins inside every cell[14]. These fields control orientation and positioning of cells, differentiation of muscle and nerve progenitor cells, as well as the proliferation rates of neoplastic (tumor) cells[1, 15]. Bioelectrical properties are important determiners of stem and progenitor cells[16, 17]. Progenitor cells differentiate into other types of cells, but are more specific than stem cells because progenitor cells can only differentiate a limited number of times. The morphogenetic field provides a working environment for the transfer of information from the cell to its targeted site in the body. The function is initiated with the production of embryonic stem cells. In mammals, there are two broad types of stem cells that differentiate into tissue: embryonic stem cells, which are isolated from the inner cell mass of blastocysts; and adult stem cells, which are found in various tissues[1]. In adult organisms, stem cells and progenitor cells act as a repair system for the body, replenishing adult tissues. In a developing embryo, stem cells can differentiate into all specialized cells and also maintain the normal turnover of regenerative organs, such as blood, skin, or intestinal tissues. From stem cells come all our body's cells that mature and eventually form body tissues and organs.

Potency specifies the differentiation potential of a stem cell[18]. Totipotent (aka omnipotent) stem cells can differentiate into embryonic and extra-embryonic cell types. Embryonic stem cells have the ability to multiply and are omnipotent, meaning they can differentiate into any cell type in the body[19]. Pluripotent stem cells are the descendants of totipotent cells and can differentiate into many but not all cells. Pluripotent means they have a predetermined and limited cell fate, thus having a limited ability to differentiate. Examples of pluripotent stem cells are embryonic stems cells older than four days, embryonic germ cells and embryonic carcinoma cells.

Multipotent stem cells can differentiate into a number of cells, but only those of a closely related family of cells. Examples include neural cells and glial (brain) or hematopoietic (blood) cells; and mesenchymal stem cell (MSC) that can differentiate into adipose tissue (fat), osteoblasts (bone), chondrocytes (cartilage) or fibroblasts (skin, ligaments and tendons), stromal cells (bone marrow), and myocytes (muscle). These tissues comprise most of the human physiology, but MSCs are limited in the number of tissues they can differentiate into. Adult stem cells are also pluripotent.

Oligopotent stem cells can differentiate into only a few cells, such as lymphoid (immune) or myeloid (bone marrow) stem cells. Stem cells that become mature blood cells are called hematopoietic stem cells (HSCs) and reside in the bone marrow. Since HSCs form red blood cells, white blood cells and platelets, they play a crucial role in the maintenance and support of our immune system. Unipotent cells only produce themselves, but have the property of self-renewal, which distinguishes them from non-stem cells such as muscle stem cells. Epithelial (skin cells) are an example of unipotent cells. What is amazing is that embryonic stem cells know which direction to send these cells in order to build the human body. The information is exchanged through our biofield so that each cell knows what every other cell is doing. Rarely does this information get distorted, but when it does a teratoma can form. A teratoma is an encapsulated tumor that can have teeth, bone and sometimes, eyes, hands, feet or limbs[20].

Bioelectric cues are important regulators of cell behavior, cell proliferation and apoptosis, as well as cell migration, cell orientation, and cell differentiation trajectory[21]. Each organ in our body is governed by bioelectric signals before it develops the cell pattern that forms its shape[22-24]. These bioelectric signals provide the energetic blueprint through which any particular body part is developed. One organ that generates significant fields affecting our entire body is the heart. Our heart gives off stronger electrical signals than any of our other organs making it the strongest electrical generator in our body[25,26]. Human brain cells also fire electrical impulses to communicate with one another, but not with the intensity of the heart. Our heart sends information to the field first then relays it to the brain[27]. Ion channels located in cell membranes work in sync to deliver messages throughout both our brain and our heart. When the electromagnetic impulses of our heart and brain synchronize, we are more intelligent, intuitive, and healthier[28].

Electrical activity is the conduit by which the heart and brain relay messages throughout the human body. Electrical impulses control the muscles which create the pumping motion in our heart. The sinus node is a section in the right upper heart chamber where the impulses are initiated. These electrical signals travel across the heart, causing the muscles to contract and pump blood throughout our body. The heart responds faster than the brain to outside information[29]. Inappropriate firing of electrical charges in our heart can result in serious medical conditions such as rapid or irregular heartbeat, skipping beats, fainting, dizziness, a feeling like our heart is jumping out of our chest, shortness of breath, and tachycardia which is a form of arrhythmia due to an abnormality in the heart's rhythm[30]. "Energy cardiology tells us that the signals produced by the heart are all of regulatory importance. The heart is constantly emitting sound in the form of pressure waves; heat, light, electrical, magnetic and electromagnetic signals at different times, because they travel at different velocities throughout the circulatory system. By far the heart generates the largest rhythmic electromagnetic signal in the body"[27]. Coherent (smooth and ordered) patterns from our heart rhythms synchronize with the brain and other systems of the body including the organs and glands to create synchronization with our heart. Incoherent (chaotic) patterns create static noise, which interferes with or blocks energy flow throughout our field. This static in our field deters healing and regeneration. Our biofield is continually supplied with information coming from the pressure waves of our heart. It can sense, feel, remember and process information that is independent from our brain[27]. Our heart has its own nervous system and with it carries an enormous amount of charge[31]. The pressure wave in the presence of this charge inside the chamber of the heart is sufficient to imprint information. If the heart is imprinting information, the cells receive this information[32]. No wonder the heart is the first organ to develop in the fetus. Our heart gives instructions for the cell assignments to generate the remaining organs.

Electromagnetic field (EMF) has been shown to disturb morphogenesis during embryonic development. Studies of histone (high alkaline proteins) synthesis in early sea urchin embryos indicated rotating 60-Hz magnetic field decreased zygotic expression of early histone genes at the morula stage suggesting this decrease in early histone production was limiting to cell proliferation[33]. Histones package and order DNA into structural units called nucleosomes[34]. EMF has also been found to reverse the energy field of a planarian flatworm with a head-tail dipole that grew despite cutting the regenerating segments. The induced reversal of the field produced reversed anterior-posterior polarity in the fragments, suggesting the planarian field can transmit morphogenetic information[35]. Planarian flatworm pieces with their original anterior end oriented toward the cathode developed normally, but pieces oriented toward the anode showed head development in the tail end, developed two heads, or underwent reversal of original polarity, depending on the current density of the field. This is an example of currents involved in both development and regeneration, since many planarian species normally reproduce by splitting in half. Our electrophysiology also provides the fields and ion currents carrying important information to regenerate limbs[36], initiate cell migration and orientation through the embryo[37], provide morphogenetic patterns with histological differentiation[38] in amphibian, avian and invertebrate models.

After our body is fully developed and functioning, morphogenesis regulates the regeneration of damaged cells and organs. The process of regeneration restores the morphology of an organism despite trauma. Regeneration is a specialized type of morphogenesis because it involves the rebuilding of existing structure amidst surrounding tissue. In replacing lost tissue (such as skin or blood) or organs (such as the liver), embryonic developmental mechanisms often need to be recruited to restore the original pattern. Shunt experiments which disturb the natural fields of organisms provide a way to test the natural currents in regeneration. Light-emitting elements in scaffolds and bioreactors that regenerate body parts have determined how and where cells will grow on scaffolds *in vitro* [39], and on regenerative sleeves used in organ regeneration *in vivo* [40]. Since electromagnetic fields already exist within organisms this gives them built-in biological patterns that form templates for regaining their function [41, 42]. Most developmental biologists are open to the idea of a holistic or integrative model of biological organization, but few have offered a mechanism for determining how the embryonic stem cells create a completely functioning organism. "As more genomes are sequenced, genes are cloned and proteins are characterized, it is not unreasonable to accept the idea that morphogenetic fields influence pattern organization in the human body"[1] says developmental biologist Michael Levin. Morphogenetic fields are not fixed – they evolve. For example, the fields that organize the activity of the nervous system are inherited through morphic resonance, indicating a collective, instinctive memory. Since all cells come from other cells, then all cells can inherit fields of organization[43]. Genes play an essential part of this organization, but they do not explain the organization itself. Some genes enable organisms to make particular proteins, while other genes are involved in the control of protein synthesis[44].

Since the human biofield is made up of standing, scalar waves that are organized, structured and contain a great deal of information[45], it is possible for these energy waves to display destructive as well as constructive interference. In physics, interference is the addition (superposition) of two or more waves resulting in a new wave pattern. Constructive interference amplifies energy while destructive interference (of equal frequency and opposite phase) annihilates it. All energy interacts with other energy and even when it appears to be static, it is not. Energy medicine amplifies our body's ability to heal itself because it harmonically affects the amplitude of the atoms in the cells of the patient, displaying constructive coherent frequencies, instead of destructive interference or noise[3]. What prevents energy medicine practitioners from causing harm is the entrainment of the practitioners' energy field with a healthy resonance (0.3 – 30 Hz - which are cycles/second) in a coherent or harmonic pattern. This resonance feels good and provides the client with a healing environment in which the body can regenerate its cells and organs.

Information flows from our biofield through our body via our nervous system and our tissues[46]. This information system functions as structured order or a matrix. Information organized in a certain order and communicated in specific directions is known as a vector. A vector field model can be used as a distribution of directions from which Shannon's entropy measures the information content in the field[47]. In information theory, entropy is a measurement of the uncertainty associated with a random variable[48]. In structured matrices information is not random but based on a system of negative entropy whereby the system does not decay but displays a tendency towards increasing order. This concept was introduced by Erwin Schrödinger in his popular-science book *What is Life – the Physical Aspect of the Living Cell*[49].

Holograms

In order to understand the holographic model of a human biofield it is important to understand how a hologram works. The phenomenon that makes holography possible is known as interference. Interference is a crisscrossing pattern of information that occurs when two or more waves run through each other. A hologram is made by sending a single laser beam through an optical device known as a beam splitter in order to create two laser beams which originate from the same source. One of the beams is called the *reference beam* and it passes through a diffusing lens that spreads it from thin orderly rays into a flared beam (see figure 1).

The reference beam is guided by mirrors to land on an unexposed photographic plate. At the same time the second beam known as the *illuminating beam,* mimics the reference beam by passing through a second diffusing lens. The difference between the two beams is that the light from the illuminating beam is used to illuminate the object being photographed, while the light from the reference beam bounces off the object and falls on

ETIOLOGY

the photographic plate. When the uninterrupted light from the illuminating beam reflects off the object being photographed and intersects the light of the reflecting beam, an interference pattern is created. The interference pattern created by the laser light and captured on the photographic file is what produces the hologram. Holographic photography is very different from photographs taken with incoherent light which is scattered compared with the coherent, organized light of a laser beam. In this same extraordinary fashion both our mind and body receive information from our biofield (referencing energy) and translate it as coherent organized illuminating energy through our cells, which eventually builds our entire human form during the morphogenetic process.

The physics of holography is well understood, but the mathematical aspects are complex, involving hyperbolic geometry[50]. A common analogy of holographic interference patterns is explained as stones being dropped into water. If one rock is dropped into water it creates a series of concentric circular waves, moving away from the center of impact. If two rocks are dropped, one after the other, the wave patterns interfere with one another. Holographic interference is created when the reference beam interferes with the illuminating beam. The photographic emulsion film captures the interference pattern and projects the subject being photographed into a 3 dimensional object and a virtual object is created from the original. Some holograms will show the object in such a way that you can move around them and see it from the top, bottom and all angles as if it were tangible. This idea is analogous to interference in the reference beam (biofield) producing interference in the illuminating beam (cell communication) creating disease. An interesting phenomenon of holography is that we can snip off a small piece of the holographic film and hold it up to laser light and still see the entire virtual object intact. Not a piece of the object, as you would see if you cut orthodox photographic film apart, but the entire object. This phenomenon mimics the holographic function of DNA. The smallest sample of DNA can reproduce the entire organism. Not just the part of the organism where the DNA was extracted, but the entire organism as a whole. The superposition principle in physics, mathematics, and engineering, explains the concept of overlapping of waves. Holographic superposition describes coherence and interference in the hologram exactly the way molecular modeling appears in the human biofield[51]. Superposition also explains how information exchange takes place in human morphogenesis[52]. The superposition model is responsible for cell, tissue and organ creation in the morphogenetic field[53]. This model provides information that is responsible for healing and regeneration when coherent, and creating disturbances in our biofield when distorted or incoherent[45].

Ribonucleic acid (RNA) uses a holographic model when it replicates DNA, because every protein in a strand of DNA contains the whole organism. Cell biologists have shown that every cell contains a copy of the master DNA blueprint[54]. Holography records the phases and amplitudes of light waves reflected from the object the same way RNA captures the frequency patterns of DNA to replicate it[55]. Transfer RNA translates the polypeptide-specifying sequences of the cell DNA the same way holography phases are recorded as interference patterns and produced by the reflected light. Each point on the hologram (RNA) receives light reflected from every part of the illuminated object (DNA) which contains the complete visual record of the object as a whole. The original laser light is representative of the DNA in our cells, and the refracting laser light is the (RNA). The holographic principles equate to recorded information (RNA) containing the energy template of the entire body. With the hologram obtained from the resonance of the polypeptide-specifying sequences of the cell DNA, a coherent set of frequencies is produced, each an exact replica of the original waves that were imbedded on the DNA when the hologram was made. RNA appears in complete three-dimensional form with highly realistic effects, just like the virtual image of a hologram. The reconstructed DNA has all the visual properties of the original object. Within this information is the composition of all body parts as they relate to three dimensional time and space. Not only do the cells know where to form a body part spatially, they also know when the time is right to form that body part. But all information cannot be read in the same time from the same angles. For instance what we perceive as a growing fetus is just a part of a quantity of different elements, or different perspectives at different positions in different time frames[56].

Every strand of DNA in every cell contains enough information to make an entire human body from scratch, but DNA is considered a blueprint because it is just a template. Theoretically you could clone the physical body of a person, but you would not be able to clone the human being; not unless the soul of the original human being reincarnated into its clone. This is because a person is not the sum of its cells (biology), molecules (chemistry) and atoms (physics). A human being has a conscious mind and emotions (psychological aspects), and also a soul (spiritual aspects). Cloning creates a physical body that looks like the original cell structure, with physical features of the original person, but the clone would likely not feel, love or react like the original person. Neuroscientists will argue that thoughts and emotions are regulated by genetic patterns in our hormones, neuropeptides, and growth factors, but energy field theory would argue that emotions come from the biofield patterns of their original generator – the human spirit in reaction to its environment, not in reaction to its cell structure.

Systems biology focuses on complex interactions in biological systems, using holism instead of reductionism, where every hormone, every neuropeptide, every cytokine and every growth factor knows what every other hormone, neuropeptide, cytokine and growth factor is doing at the same time[57, 58]. Disease begins with disorganized frequencies of incoherent information which can stem from unhealthy eating habits, unhealthy thought patterns, or toxic environments that interrupt the communication exchange of healthy cells, interfere with cell signaling patterns, and cause chaos or stress on the system. Healing is the simultaneous exchange of constructive interference patterns in a holographic pattern. From conception our body uses its morphogenetic field to maintain the form and shape of our body, aiding it in recovery from disease or damage, promoting healing and regeneration. From embryonic stem cells to fully developed organisms, we rely on morphogenetic fields functioning in a holographic model to create and regenerate healthy human bodies.

References

1. Levin, M., *Bioelectric mechanisms in regeneration: unique aspects and furture perspectives.* Semin Cell Dev Biol, 2009. **20**(5): p. 543-556.
2. Levin, M., *Bioelectromagnetics in Morphogenesis.* Bioelectromagnetics, 2003. **24**: p. 295-315.
3. Hunt, V., *The promise of bioenergy fields: an end to all disease.* http://www.spiritofmaat. com/archive/nov1/vh.htm, 2012 retrieved May 15.
4. Alberts, B., et al., *Universal Mechanisms of Animal Development.* Molecular Biology of the Cell, 2002. **4th ed**(Garland): p. ISBN 0815332181.
5. Jacobson, A. and A. Sater, *Features of embryonic induction.* Development, 1988. **104**(3): p. 341–359.
6. Gilbert, S., J. Opitz, and R. Raff, *Resynthesizing evolutionary and developmental biology.* Dev Biol., 1996. **173** (2): p. 357–372.
7. Gilbert, S., *Developmental biology.* 2003. **7th ed**(Sinauer Associates: Sunderland, Mass): p. 65–66.
8. Turing, A., *The chemical basis of morphogenesis.* Philosophicals Transactions of the Royal Society of London, 1952. **B 237**(641): p. 37–72.
9. Levin, M., *The wisdom of the body: future techniques and approaches to morphogenetic*

fields in regenerative medicine, developmental biology and cancer. Regen Med., 2011. **6**(6): p. 667-73.

10. Szent-Gyorgyi, A., *Introduction to a submolecular biology.* Academic Press, 1960. **New York**.

11. Sheldrake, R., *A New Science of Life: The Hypothesis of Formative Causation.* Los Angeles: JP Tarcher, 1981: p. 52.

12. Tsonis, P., *Effects of carcinogens on regerating and non-regenerating limbs in amphibia (review).* Anticancer Res, 1983. **3**(3): p. 195-202.

13. Brockes, J., *Regeneration and cancer.* Biophys Acta, 1998. **1377**(1): p. M1-M11.

14. Adams, D., Levin, M, *General principles for measuring resting membrane potential and ion concentration using fluorescent bioelectricity reporters.* Cold Spring Harb Protoc., 2012. **pii: pdb.top067710. doi: 10.1101/pdb.top067710**(4).

15. McCaig, C., Song, B, Rajnicek, A, *Electrical dimensions in cell science.* J Cell Sci., 2009. **122**(Pt 23): p. 4267-4276.

16. Root, C., Velazquez-Ulloa, N, Monsalve, G, Minakova, E, Spitzer, N, *Embryonically expressed GABA and glutamate drive elctrical activity regulating neurotransmitter specificaiton.* J Neurosci, 2008. **28**('8): p. 4777-4784.

17. Ng, S., Chin, C, Lau, Y, *Role of voltage-gated potassium channels in the fate determination of embryonic stem cells.* J Cell Physiol, 2010. **224**(1): p. 165-177.

18. Schöler, H., *The Potential of Stem Cells: An Inventory.* Humanbiotechnology as Social Challenge, 2007. **In Nikolaus Knoepffler, Dagmar Schipanski, and Stefan Lorenz Sorgner**(Ashgate Publishing, Ltd).

19. Mitalipov, S., Wolf, D, *Totipotency, pluripotency and nuclear reprogramming.* Adv. Biochem. Eng. Biotechnol, 2009. **114**: p. 185–99.

20. Chi, J., Lee, Y, Park, Y, Chang, K, *Fetus-in-fetu: report of a case.* American Journal of Clinical Pathology, 1984. **82**(1): p. 115–119.

21. Levin, M., *Molecular bioelectricity in developmental biology: New tools and recent discoveries.* Bioessays, 2012. **34**: p. 205-217.

22. Nishiyama, M., von Schimmelmann, M, Togashi, K, Findley, W, *Membrane potential shifts caused by diffusible guidance signals direct growth-cone turning.* Nat Neurosci, 2008. **11**: p. 762-771.

23. Ozkucur, n., Perike,S, Sharma, P, Funk, R, *Persistent directional cell migration requires ion transport proteins as direction sensors and membrane potential differences in order to maintain directedness.* BMC Cell Biol, 2011. **12**: p. 4.

24. Cao, L., Pu, J. Zhao, M, *GSK-3beta is essential for physiological electric field-directed Golgi polarization an doptimal electrotaxis.* Cell Mol Life Sci, 2011. **68**: p. 3081-93.

25. Willis, H., *Naming of the Waves in the ECG, With a Brief Account of Their Genesis.* Circulation (suppl I), 1998. **98**(18).

26. Berger, R., Akselrod, S, Gordon, D, Cohen, R, *An efficient algorithm for spectral analysis of heart rate variability.* IEEE Trans Biomed Eng, 1986. **33**: p. 900–904.

27. McCraty, R., Atkinson, M, Tomasino, D, Bradley, R, *The Coherent Heart: Heart-Brain Interactions, Psychophysiological Coherence, and the Emergence of System-Wide Order* HeartMath Research Center, Institute of HeartMath, 2006. **Publication No. 06-022**(Boulder Creek, CA).

28. Chopra, D., *Quantum Healing: Exploring the Frontiers of Mind/Body Medicine.* Bantam Books, New York, 1989: p. 21.
29. Oschman, J., *Energy and the healing response.* J Bodywork Movement Therap, 2005. **9**: p. 3-15.
30. Tortora, G. and S. Grabowski, *Principles of Anatomy and Physiology.* Wiley & Sons. Brisbane, Singapore & Chichester, 2000.
31. Sater, A., Jacobson, A, *The restriction of the heart morphgenetic field in xenopus laevis.* Dev Biol., 1990. **140**(2): p. 328-336.
32. Dehaan, R., *Morphogenesis of the vertebrate heart.* Organogenesis, Holt, Rinehart and Winston: New York, 1965: p. 377-419.
33. Cameron, I., et al., *Environmental magnetic fields: Influences on early embryogenesis.* J Cell Biochem., 1993. **51**(417-425).
34. Cox, M., Nelson, D, Lehninger, A, *Lehninger Principles of Biochemistry.* San Francisco: W.H. Freeman, 2005(ISBN 0-7167-4339-6).
35. Levin, M., *Heads or Tails: Cells' Electricity Decides What to Regenerate.* Journal Chemistry & Biology, 2011.
36. Jenkins, L., Duerstock, B, Borgens, R, *Reduction of the current of injury leaving the amputation inhibits limb regeneration in the red spotted newt.* Dev Biol., 1996. **178**: p. 251-262.
37. Shi, R., Borgens, R, *Three-dimensional gradients of voltage during development of the nervous system as invisible coordinates for the establishment of embryonic pattern.* Dev Dyn, 1995. **202**: p. 101-114.
38. Borgens, R., Shi, R, *Uncoupling histogenesis from morphogenesis in the vertebrate embryo by collapse of the transeural tube potential.* Dev Dyn, 1995. **203**: p. 456-467.
39. Stroh, A., Tsai, H, Ping Wang, L, *Tracking stem cell differentiation in the setting of automated optogenetic stimulation.* Stem Cells, 2010. **29**(1): p. 78-88.
40. Hechavarria, D., Devilde, A, Braunhut, S, Levin, M, Kaplan, D, *BioDone regenerative sleeve for biochemical and biophysical stimulation of tissue regeneration.* Med Eng Phys, 2010. **32**(9): p. 1065-1073.
41. Brockes, J., *Amphibian limb regeneration: rebuilding a complex structure.* Science, 1997. **276**: p. 81-87.
42. Keller, R., *The origin and morphogenesis of amphibian somites.* Curr Top Dev Biol. , 2000. **47**: p. 183-246.
43. Sheldrake, R., *Mind, Memory, and Archetype Morphic Resonance and the Collective Unconscious.* Psychological Perspectives, 1987. **18**(1): p. 9-25.
44. Vimalra, j.S., Selvamurugan, N, *MicroRNAs: Synthesis, Gene Regulation and Osteoblast Differentiation.* Curr Issues Mol Biol, 2012. **15**(1): p. 7-18.
45. Hunt, V., *Infinite Mind: Science of the Human Vibrations of Consciousness.* Malibu Publishing Company, 1996. **Malibu, CA.**
46. McCulloch, W., Pitts, W, *A logical calculus of the ideas immanent in nervous activity:* Bull. Math. Biophysics, 1943. **5**(4): p. 115–133.
47. Lipton, B., *The Biology of Belief: Unleashing the power of consciousness, matter and miracles.* Elite Books, 2005. **Santa Rose, CA.**
48. Shunsuke, I., *Information theory for continuous systems.* World Scientific, 1993: p. 2.

49. Schrödinger, E., *What is Life - the Physical Aspect of the Living Cell*. Cambridge University Press, 1944.

50. Kasper, J. and S. Feller, *The complete book of holograms: how they work and how to make them*. John Wiley and Sons, Inc.: New York, 2001.

51. Jossa, U., Spahnib, H, *Holographic superposition of molecular models*. Journal of Molecular Graphics, 1986. **4**(3): p. 143–144.

52. Maini, P., *Superposition of modes in a caricature of a model for morphogenesis*. Journal of Mathematical Biology, 1990. **28**(3): p. 307-315.

53. Nagorcka, B.N., Manorangan, V. S., Murray, J. D, *Complex spatial patterns from tissue interactions--an illustrative model*. J. Theor. Biol, 1987. **128**: p. 93-112.

54. Copley, J., *Proof of Life*. New Scientist, 2003. **177**(2383): p. 28-31.

55. Kaiser, F., Josse, A, Kornberg, A, *Analysis of nearest neighbor base frequencies in the RNA of a mammalian virus: encephalomyocarditis virus*. J . gen. ViroL, 1968. **2**: p. 469-472.

56. Gerber, R., *Vibrational Medicine*. Bear & Company, Santa Fe, NM, 1998.

57. http://www.systemsbiology.org/Intro_to_Systems_Biology/Systems_Biology_--_the_21st_Century_Science., *Systems Biology: the 21st Century Science"*. Institute for Systems Biology. Retrieved 15 January 2011, 2011.

58. Mesarovic, M., *Systems Theory and Biology*. Berlin: Springer-Verlag, 1968.

CHAPTER THIRTEEN
How does my Physical Health affect my Emotional Health and Vice Versa?

Our physical health can profoundly affect our emotional health. People having difficulty coping with a chronic illness are more likely to become depressed[1]. Thoughts and feelings generated in our mind can influence the release of hormones in our endocrine system which affects the cells in our body. Clinically depressed people often have physical symptoms of constipation, changes in appetite, insomnia or lethargy[2]. The visceral reaction of hearing a diagnosis of cancer can send a patient into a state of emotional shock. The same can be said of an HIV/AIDs diagnosis or any other life threatening disease. Emotional and psychological health issues can be brought on by suffering the loss of a loved one, juggling the responsibility of a busy lifestyle, caregiving an older family member, dealing with Alzheimer's disease, or dealing with alcoholism or drug abuse. People often experience back pain, joint stiffness, and tense muscles when anxiety goes unresolved. Dangerous physical effects can be experienced in response to emotional blockages and stress. These include high blood pressure, rapid heartbeat (tachycardia), irregular heartbeat (arrhythmia), or chest pains–all of which can raise the risk of cardiac disease or stroke[3]. Emotional health can be affected by physical pain as well. Patients with severe chronic pain can become suicidal[1].

Research shows our digestive system is capable of creating sadness, grief and trauma, loss of appetite, overeating, dizziness, nausea, constipation or diarrhea[4]. Many people also experience tension headaches from unresolved or unidentified stress. Others experience exhaustion, insomnia, and dry mouth, shortness of breath, respiratory issues or profuse sweating in response to emotional trauma or blockages. Our immune system becomes compromised as emotional stress makes us more susceptible to disease and illness[5]. Researchers have been studying the role of psychosocial factors in the prevention of heart disease and found that depression, social isolation, chronic anger and anxiety can affect not only our behavior but our physiology[6]. Depression can also lead to stroke[7]. People who are experiencing depression and anxiety are more likely to live unhealthy lifestyles, including over-eating, lack of exercise, increased smoking and alcohol consumption. These behaviors can also lead to cardiovascular disease and plaque rupture, which is the main cause of heart attack[8, 9].

Our emotional body affects our physical body

Our body responds to the way we think, act and feel. This is known as the mind-body connection. When we are stressed, anxious or upset, our body gives off signals. For example, high blood pressure or stomach ulcers can develop after a stressful event such as the death of a loved one. Low back pain is a physical sign that emotional health is out of balance[10]. Chronic pain can interfere with our memory[11]. Chronic pain can also have a profound effect on mood because it interferes with every aspect of life, such as appetite and sleep. People who are in constant pain may worry they cannot work or go about their daily activities. Both incurable illness[12] and chronic pain[13] can cause depression. Pain is more than just unpleasant sensations in our nervous system; it involves our perceptions, feelings, and thoughts. The worse we anticipate our pain is going to be, the worse it becomes[14].

The physical and emotional toll of living in constant pain leads nearly a third of people with chronic pain to become clinically depressed. About 75 percent of people who are being treated for depression report physical symptoms, including pain[15]. If pain can lead to emotional distress, the reverse is also true. The more trouble we have dealing with stress, the more likely we are to experience physical pain. Research studies report patients who were under mental distress or who had chronic pain (not in the lower back) were three times more likely to develop low back pain than those who had better coping skills[16]. Stress and pain can cause muscle tension causing more pain. Fear and avoidance of pain will cause people to become inactive weakening the body even further[17].

Poor emotional health weakens the body's immune system[18]. Poor emotional health decreases our desire to exercise, eat healthy food, and increases consumption of alcohol, tobacco and drugs. The amygdala is the part of the brain responsible for emotions. It is in the temporal lobe of the brain, which is in front of the hippocampus, and deals with fear conditioning. Anxiety, autism, depression, post-traumatic stress disorder, and phobias are all suspected of being linked to abnormal functioning of the amygdala[19]. Scientists are studying how the amygdala might affect people who suffer from bipolar illnesses[20], because it is linked to both fear responses and pleasure.

The hippocampus is the part of the brain that is essential to memory. Its structure is significantly smaller in veterans with post-traumatic stress disorder than in those without the condition. Using magnetic resonance imaging (MRI), researchers scanned the brains of 36 veterans (17 with combat-related PTSD and 19 without) and found that the hippocampus was more than 11 percent smaller on average in the veterans with PTSD[21]. The hippocampus is located in the medial temporal lobe of the brain, and its function is to convert short-term memory to long-term memory. The more severe the PTSD symptoms are, the smaller the hippocampus is.

The diagnosis of an autoimmune disease can be as emotionally challenging as it is physically demanding. People with autoimmune disease often report feeling stressed, overwhelmed and emotionally drained. Sometimes stress and other emotional problems can trigger the symptoms of autoimmune disease causing it to get worse. Emotional symptoms of the following autoimmune diseases include irritability and insomnia with Grave's disease; and fatigue and depression in Hashimoto's thyroiditis and Rheumatoid arthritis. A diagnosis of chronic illness can cause anger, denial, fear, frustration, depression, anxiety and isolation[22].

The emotional challenges of a terminal illness

People diagnosed with incurable disease are often plagued with anxiety about death, the emotional welfare of family members, financial concerns, and whether they will be a burden on others. The mental, emotional and psychological processes that arise from having a terminal illness are complex. Treatments can leave the patient struggling with a poor quality of life, disabling them physically as well as emotionally. Immune compromised people are often quarantined from society to avoid getting communicable diseases. Patients with chronic conditions experiencing changes in their employment status have even more anxiety and distress. The proportion of patients with diabetes or rheumatoid arthritis who have a psychological disorder runs between 20 and 25 percent[23]. Depression and helplessness can undermine a patient's ability to cope with pain and cause additional stress on family relationships[24]. Challenges stemming from terminal illness prove to be intense in all areas of life. Families change, social networks change, mental and emotional attitudes about life change as death becomes a reality. Patient education, support groups, and cognitive behavioral therapies, are reported to have beneficial effects on health and emotional well-being and are significantly higher than improvements attained with usual medical care alone (i.e., pharmaceuticals or surgery)[25]. Hospice and palliative care programs help alleviate suffering and prepare patients for end of life care (see http://iweb. nhpco.org for information on palliative care). There are also support groups for family members caring for palliative relatives. Whether you are a patient or caregiver support is important. As life expectancy increases and medical treatments advance, more people will become caregivers. Caregiving involves changes in the family dynamic, household disruption, financial pressure, and an increased amount of work that all create added stress on caregivers with little reward for a happy outcome. Due to these lifestyle changes it is important to take time to focus on our stress level and emotional stability so they do not cause good health to go bad. Depression, extreme lifestyle change and burn out can all lead to fibromyalgia[26].

There are several self-help and alternative medical treatments available to prevent physical health problems from affecting us emotionally and emotional health problems from affecting us physically. Energy psychology and energy medicine work with the mind-body connection to balance distorted energy in our biofield preventing further complication. Energy Psychology (EP) is a "family of evidence supported modalities that balance, restore and improve human functioning by combining physical interventions (using the acupuncture system, the chakras and other ancient systems of healing) with modern cognitive interventions such as imagery-based exposure therapy" (www.energypsych.org). Other energy medicine (EM) treatments serve as both a complement or alternative to medical care and provide a complete system for self-care and self-help. EM addresses physical illness and emotional or mental disorders, and also promotes high-level wellness and peak performance. It involves deep breathing techniques along with meditation, both helpful in regulating emotion associated with chronic and terminal illness. People with good emotional health are aware of their own thoughts and feelings. Good emotional health means finding ways of coping with stress and problems that are a normal part of life. Emotional Freedom Technique (EFT) has been reported to clear emotions and limiting beliefs that can lead to extreme anxiety[27] as well as reduce the symptoms of fibromyalgia[28]. EFT can also clear emotional issues surrounding expectations and outcomes about our present and future[29]. EFT helps to increase self esteem and promote healthy relationships. It has also shown to be effective in treating post traumatic stress disorder[30].

It is important to be aware of both good stress and bad stress. A new job or job promotion, getting married, a new baby, a new house are all exciting, but can create as much added stress and anxiety as getting laid off from a job, having a child return home, getting divorced, bankruptcy, suffering an illness or injury, or the death of a loved one. It is important to watch for signs of stress which include shallow breathing, muscle tension, headaches, changes in eating habits and lack of sleep. Lack of resilience was found to be a leading predictor in how people cope with physical and emotional stressors. Spirituality, purpose in life, and trait anxiety contribute to different levels of resilience in patients with depression and/or anxiety disorders. Low spirituality was revealed as a leading predictor of lower-resilience groups[31]. Low purpose in life and less frequent exercise were associated with the low- and medium-resilience groups, respectively. Severe trait anxiety was characterized with the low- and medium-resilience groups. The ability to refrain from focusing on problems along with the ability to look for solutions create a much higher quality of life in people suffering from both physical and emotional health issues. Emotional illness is often more difficult to detect than physical illness. It is important to watch for signs of emotional distress so they do not lead to physical illness and vice versa. Energy psychology along with other energy medicine therapies, treat both at the same time.

References

1. Barkin, R., Barkin, S, Irving, G, Gordon, A, *Management of chronic noncancer pain in depressed patients.* Postgrad Med, 2011. **123**(5): p. 143-154.
2. Pattanayak, R., Sagar, R, *Depression in dementia patients: issues and challenges for a physician.* J Assoc Physicians India, 2011. **59**: p. 650-52.
3. Villemure, C., Bushnell, M *Cognitive modulation of pain: how do attention and emotion influence pain processing.* Pain, 2002. **95**: p. 195–199.
4. Kiecolt-Glaser, J., *Stress, food, and inflammation: psychoneuroimmunology and nutrition at the cutting edge.* Psychosom Med, 2010. **72**(4): p. 365-369.
5. Reiche, E., Nunes, S, Morimoto, H, *Stress, depression, the immune system, and cancer.* Lancet Oncol, 2004. **5**(10): p. 617-25.
6. Merswolken, M., Deter, H, Siebenhuener, S, Orth-Gomér, K, Weber, CS. , *Anxiety as Predictor of the Cortisol Awakening Response in Patients with Coronary Heart Disease.* Int J Behav Med, 2012.

7. Naess, H., Lunde, L, Brogger, J, *The Triad of Pain, Fatigue and Depression in Ischemic Stroke Patients: The Bergen Stroke Study.* Cerebrovasc Dis, 2010. **33**(5): p. 461-465.

8. Mehrzad, R., Spodick, D, *Pericardial Involvement in Diseases of the Heart and Other Contiguous Structures: Part I - Pericardial Involvement in Infarct Pericarditis and Pericardial Involvement following Myocardial Infarction.* Cardiology, 2012. **121**(3): p. 164-176.

9. Mehrzad, R., Spodick, D, *Pericardial Involvement in Diseases of the Heart and Other Contiguous Structures: Part II - Pericardial Involvement in Noncardiac Contiguous Disorders.* Cardiology, 2012. **121**(3): p. 177-183.

10. Edit, V., Eva, S, Maria, K, Istvan, R, Agnes, C, Zsolt, N, Eva, P, Laszlo, H, Peter, T, Emese, K, Klara, T, Gyula, P, *Psychosocial, educational, and somatic factors in chronic nonspecific low back pain.* Rheumatol Int, 2012. **Apr 3. [Epub ahead of print]**.

11. Denk, F., McMahon, S, *Chronic pain: emerging evidence for the involvement of epigenetics.* Neuron, 2012. **73**(3): p. 435-444.

12. Mayr, M., Schmid, R, *Depression in pancreatic cancer: sense of impending doom.* Digestion, 2010. **82**(1): p. 1-3.

13. Meeus, M., Nijs, J, Van Mol, E, Truijen, S, De Meirleir, K, *Role of psychological aspects in both chronic pain and in daily functioning in chronic fatigue syndrome: a prospective longitudinal study.* Clin Rheumatol, 2012 Jun;31(6):921-9.

14. Caes, L., Uzieblo, K, Crombez, G, De Ruddere, L, Vervoort ,T, Goubert, L, *Negative emotional responses elicited by the anticipation of pain in others: psychophysiological evidence.* J Pain., 2012. **13**(5): p. 467-76.

15. Rouwette, T., Vanelderen, P, Roubos, E, Kozicz, T, Vissers, K, *The amygdala, a relay station for switching on and off pain.* Eur J Pain, 2012 Jul;16(6):782-92.

16. Schneider, S., Junghaenel, D, Keefe, F, Schwartz, J, Stone, A, Broderick, J, *Individual differences in the day-to-day variability of pain, fatigue, and well-being in patients with rheumatic disease: Associations with psychological variables.* Pain, 2012 Apr;153(4):813-22.

17. Moldovan, A., *Emotional Aspects of Low Back Pain.* Journal of Cognitive and Behavioral Therapies, 2009. **9**: p. 83-93.

18. Zhou, F., Zhang, W, Wei, Y, Xu, K, Hui, L, Wang, X, L,i M, *Impact of comorbid anxiety and depression on quality of life and cellular immunity changes in patients with digestive tract cancers.* World J Gastroenterol, 2005. **11**(15): p. 2313-2318.

19. Iidaka, T., Miyakoshi, M, Harada, T, Nakai, T, *White matter connectivity between superior temporal sulcus and amygdala is associated with autistic trait in healthy humans.* Neurosci Lett, 2012. **510**(154-158).

20. Linke, J., King, A, Rietschel, M, Strohmaier, J, Hennerici, M, Gass, A, Meyer-Lindenberg, A, Wessa, M, *Increased Medial Orbitofrontal and Amygdala Activation: Evidence for a Systems-Level Endophenotype of Bipolar I Disorder.* Am J Psychiatry., 2012. Mar;169(3):316-25.

21. Wang, Z., Neylan, T, Mueller, S, Lenoci, M, Trura,n D, Marmar, C, Weiner, M, Schuff, N, *Magnetic resonance imaging of hippocampal subfields in posttraumatic stress disorder.* Arch Gen Psychiatry, 2010. **67**(3): p. 296-303.

22. Walker, J., Graff, L, Dutz, J, Bernstein, C, *Psychiatric disorders in patients with immune-*

mediated inflammatory diseases: prevalence, association with disease activity, and overall patient well-being. J Rheumatol Suppl. , 2011. **88**(31-35).

23. Guthrie, E., *Emotional disorder in chronic illness: psychotherapeutic interventions.* Br J Psychiatry, 1996. **168**: p. 265-273.

24. Breitbart, W., *Identifying patients at risk for and treatment of major psychiatric complications of cancer.* Support Care Cancer, 1995. **3**: p. 45-60.

25. Martire, L., Schulz, R, *Involving Family in Psychosocial Interventions for Chronic Illness.* Curr Direc Psych Sci, 2007. **16**(2): p. 90-94.

26. Van Houdenhove, B., Luyten, P, *Customizing treatment of chronic fatigue syndrome and fibromyalgia: the role of perpetuating factors.* Psychosomatics, 2008. **49**(6): p. 470-477.

27. Benor, D., Ledger, K, Toussaint, L, Hett, G, Zaccaro, D, *Pilot study of emotional freedom technique (EFT), wholistic hybrid derived from EMDR and EFT (WHEE) and cognitive behavioral therapy (CBT) for treatment of test anxiety in university students.* Explore, 2009. **5**(6): p. 1.

28. Brattberg, G., *Self-administered emotional freedom techniques (EFT) in individuals with fibromyalgia: a randomized trial.* Integrative Medicine: a clinicians journal, 2008. **august/september**.

29. Palmer-Hoffman, J., Brooks, A, *Psychological Symptom Change after Group Application of Emotional Freedom Techniques.* Energy Psychology: Theory, Research, & Treatment, 2011. **3**(1): p. 57-72.

30. Church, D., *The treatment of combat trauma in veterans using EFT (Emotional Freedom Techniques): a pilot protocol.* Traumatology, 2009. **15**: p. 1.

31. Min, J., Jung, Y, Kim, D, Yim, H, Kim, J, Kim, T, Lee, C, Lee, C, Chae, J, *Characteristics associated with low resilience in patients with depression and/or anxiety disorders.* Qual Life Res, 2012. **Apr 7. [Epub ahead of print]**.

CHAPTER FOURTEEN
How Do My Thoughts and Emotions Affect My Health?

The Mind-Body Connection

For over a century modern medicine has been interested in the relationship between psychiatric symptoms and immune function known as mind-body medicine. This relationship is also known as psychoneuroimmunology (PNI). PNI studies how our mind affects our nervous system which affects our immune system[1]. *Psycho* refers to our thoughts, emotions and mood states; *neuro* refers to our sympathetic, parasympathetic and central nervous systems; and *immunology* refers to our cellular structures and immune system. PNI incorporates psychology, neuroscience, immunology, physiology, pharmacology, molecular biology, psychiatry, behavioral medicine, infectious diseases, endocrinology, and rheumatology. Research suggests our mind and body communicate with each other in a bidirectional flow of hormones, neuropeptides and cytokines[2, 3]. PNI studies exact mechanisms through which specific brain immunity effects are achieved. Evidence for nervous system–immune system interactions exists on several biological levels. The immune system and the brain communicate with each other through signaling pathways[4]. The brain and the immune system are the two major adaptive systems of the body linking the hypothalamic-pituitary-adrenal axis (HPA axis) and the sympathetic nervous system (SNS). The activation of SNS during an immune response is triggered to localize the inflammatory response [5, 6]. The HPA axis responds to physical and mental challenges in order to maintain stability in part by controlling the body's cortisol level. Imbalances in the HPA axis are the cause of many stress-related diseases[7]. HPA axis activity is linked by inflammatory cytokines that stimulate adrenocorticotropic hormone (ACTH) and cortisol secretion, while, glucocorticoids suppress proinflammatory cytokines. Cytokine regulation of hypothalamic function is an active area of research for the treatment of anxiety-related disorders[8]. Cytokines mediate and control immune and inflammatory responses. Complex interactions between cytokines, inflammation and the adaptive responses maintain homeostasis in the body to protect against disease. Like the stress response, inflammatory reaction is crucial for survival. Without inflammation our bodies could not fight off the pathogens that cause disease. Studies show pro-inflammatory cytokine processes take place during depression, mania, bipolar disease, and schizophrenia, as well as autoimmune hypersensitivity and chronic infections[9]. Neuroinflammation and neuroimmune activation have been shown to play a role in the etiology of a variety of neurodegenerative disorders such as Parkinson's disease[10], Alzheimer's disease[11], schizophrenia[12], chronic pain[13], and AIDS[14].

The melanocortin system (neurons and brain stem) is the central core of emotional stress-induced anxiety, anorexia and activation of the hypothalamopituitary-adrenal (HPA) axis[15]. The mid brain or medial amygdala is highly sensitive to emotional stress. Activation of the amygdala has been shown to effect anxious behavior, food intake and corticosterone secretion[16]. Neurons in the amygdala activated by acute stress can cause depletion or increase in appetite. Stressed-induced eating was reported to be the key indicator of increased eating of sweet fatty foods in a study of emotional eaters[17], where both males and females were found to change eating habits under acute stress conditions. Stress can compromise the health of susceptible people through stress-related changes in food choices. Severe stress can increase the mortality of partners after the death of a spouse[18]. Stressful events trigger cognitive responses which induce sympathetic nervous system and endocrine changes impairing immune function[19]. Stress increases the number of white blood cells in our body, as well as decreases the numbers of helper T-cells, suppressor T-cells, and cytotoxic T-cells, B cells, and natural killer cells[20].

In her book *Molecules of Emotion*, Candace Pert explains the science behind the brain's opiate receptors and other evidence of the intimate connections between mind and body[21]. Pert's research shows that people have a difficult time discriminating between physical, mental and emotional pain. Research has been conducted linking physical pain with depression and anxiety[22]. Metabolic markers of depression have been found in the urine of rats. The metabolic profiling of urine samples from depressed rats showed disease biomarkers and pathology of depression[23].

Studies also show levels of interferon gamma (INF-γ), which can affect brain function and induce depression, are lower in high-happiness groups of people compared with unhappy people[24]. A negative correlation was found between the levels of perceived happiness and INF-γ and experimentally induced happiness, which can reduce INF-γ levels[25]. Results revealed an association between the perception of happiness and systemic inflammation[25]. It appears that happiness in an immune booster. Being happy helps fight off infection. Our thoughts can also profoundly influence our health by shutting down our immune system when we are frightened, making us less intelligent but more able to react instinctively rather than think things through before fighting or trying to escape the situation[26]. Our sympathetic nervous system (SNS) is designed to deal with acute situations, not chronic ones. Stressful environments promote sympathetic nervous system (fight or flight) activation, while serene environments encourage the regeneration of our body through our parasympathetic nervous system (rest and digest). Serenity promotes health and wellness while stressful environments promote illness and disease. The difference occurs in the stress level of the organism. For the same reason that stress expressed through our sympathetic nervous system can be detrimental to our health and well-being, calmness expressed in our parasympathetic nervous system can regenerate the vital elements of our body. Through balancing the energy of our mind, emotions and spirit, our body is able to keep the immune system strong and able to fight off disease.

In order to avoid staying in chronic SNS mode, we must monitor our reaction to environmental stressors. Fear and anxiety create interference between cell signaling molecules, increasing proinflammatory cytokines that contribute to depression, disease, fatigue and cognitive impairment[27]. If our parasympathetic nervous system cannot recover from the fear of our environment, there is increased potential for cell miscommunication and the onset of disease.

Changing our thoughts to affect our health

As we think our thoughts and feel our feelings, our bodies respond with a complex array of changes. Each thought or feeling releases a particular cascade of biochemicals in our organs[28]. Each experience triggers genetic changes in our cells by affecting our cells' membrane, cytoplasm, and nucleus. Mind-body specialists are typically psychologists or other mental health professionals who examine the affects of our mind, thoughts, attitudes and beliefs on physical health and well-being[29, 30]. Specialists use a variety of techniques such as deep breathing, guided imagery, relaxation therapy, meditation and yoga. These techniques induce oxytocin[31-33]. Oxytocin is a neuropeptide involved in a wide variety of social behaviors in diverse species. Recent research on its effects in humans has generated much interest in its role in the dynamic function of the social brain[34]. Our brain naturally produces oxytocin when we engage in activities like cuddling, hugging and having sex. Oxytocin has been shown to dramatically improve symptoms of sexual dysfunction[35] as well as reduce symptoms of schizophrenia patients[36]. It was also found to have an anti-inflammatory affect[37].

Research suggests that addressing psychological and spiritual health, which includes changes in mood, attitudes, self-image and outlook, can help in the recovery process from cancer and other forms of disease[38]. Medical treatment for incurable disease can be physically demanding and mentally exhausting[39]. Mind-body medicine, like guided imagery and relaxation, has been shown to reduce symptoms of chronic and terminal illness[40]. Mind-body techniques are important tools in achieving peace of mind, well-being, and an improved quality of life during challenging health issues.

References

1. Irwin, M., Vedhara, K, *Human Psychoneuroimmunology.* Oxford University Press:London, 2005.
2. Levy, D., *Endogenous Mechanisms Underlying the Activation and Sensitization of Meningeal Nociceptors: The Role of Immuno-Vascular Interactions and Cortical Spreading Depression.* Curr Pain Headache Rep, 2012 Jun;16(3):270-7.
3. Wei, H., Chadman, K, McCloskey, D, Sheikh A, Malik, M, Brown, W, Li, X, *Brain IL-6 elevation causes neuronal circuitry imbalances and mediates autism-like behaviors.* Biochim Biophys Acta, 2012 Jun;1822(6):831-42.
4. Watkins, A., *Psychoneuroimmunology.* Mind-Body Medicine: A clinician's guide to psychoneuroimmunology, 1997: p. 3-18.
5. Lamkin, D., Sloan, E, Patel, A, Chiang, B, Pimentel, M, Ma, J, Arevalo, J, Morizono, K, Cole, S, *Chronic stress enhances progression of acute lymphoblastic leukemia via β-adrenergic signaling.* Brain Behav Immun, 2012 May;26(4):635-41.
6. Vlcek, M., Rovensky, J, Eisenhofer, G, Radikova, Z, Penesova, A, Kerlik, J, Imrich, R, *Autonomic Nervous System Function in Rheumatoid Arthritis.* Cell Mol Neurobiol., 2012 Jul;32(5):897-901.
7. Rohleder, N., *Acute and chronic stress induced changes in sensitivity of peripheral inflammatory pathways to the signals of multiple stress systems - 2011 Curt Richter Award Winner.* Psychoneuroendocrinology, 2012 **37**(3): p. 307-316.
8. Player, M., Peterson, L, *Anxiety disorders, hypertension, and cardiovascular risk: a review.* Int J Psychiatry Med, 2011. **41**(4): p. 365-377.
9. Brietzke, E., Stertz, L, Fernandes, B, Kauer-Sant'anna, M, Mascarenhas, M, Escosteguy,

Vargas, A, Chies, J, Kapczinski, F, *Comparison of cytokine levels in depressed, manic and euthymic patients with bipolar disorder.* J Affect Disord, 2009. **116**(3): p. 214-217.

10. Rees, K., Stowe, R, Patel, S, Ives, N, Breen, K, Clarke, C, Ben-Shlomo, Y, *Non-steroidal anti-inflammatory drugs as disease-modifying agents for Parkinson's disease: evidence from observational studies.* Cochrane Database Syst Rev, 2011. **9**(11): p. CD008454.

11. Rosales-Corral, S., Acuña-Castroviejo, D, Coto-Montes, A, Boga, J, Manchester, L, Fuentes-Broto, L, Korkmaz, A, Ma, S, Tan, D, Reiter, R, *Alzheimer's disease: pathological mechanisms and the beneficial role of melatonin.* J Pineal Res, 2012. **52**(2): p. 167-202.

12. Howes, O., Fusar-Poli, P, Bloomfield, M, Selvaraj, S, McGuire, P, *From the prodrome to chronic schizophrenia: the neurobiology underlying psychotic symptoms and cognitive impairments.* Curr Pharm Des, 2012. **18**(4): p. 459-465.

13. Borsook, D., *Neurological diseases and pain.* Brain Behav Immun, 2012. **135**(Pt 2): p. 320-344.

14. Acharjee, S., Zhu, Y, Maingat, F, Pardo, C, Ballanyi, K, Hollenberg, M, Power, C, *Proteinase-activated receptor-1 mediates dorsal root ganglion neuronal degeneration in HIV/AIDS.* Brain., 2011 **134**(Pt 11): p. 3209-3221.

15. Liu, J., Garza, J, Li, W, Lu, X, *Melanocortin-4 receptor in the medial amygdala regulates emotional stress-induced anxiety-like behaviour, anorexia and corticosterone secretion.* Int J Neuropsychopharmacol, 2011. **16**: p. 1-16.

16. Sripada, R., King, A, Garfinkel, S, Wang, X, Sripada, C, Welsh, R, Liberzon, I, *Altered resting-state amygdala functional connectivity in men with posttraumatic stress disorder.* J Psychiatry Neurosci, 2012. **37**(2): p. 110069.

17. Oliver, G., Wardle, J, Gibson, EL, *Stress and Food Choice: A Laboratory Study* Psychosomatic Medicine, 2000. **62**(6): p. 853-865.

18. Boyle, P., Feng, Z, Raab, G, *Does widowhood increase mortality risk?: testing for selection effects by comparing causes of spousal death.* Epidemiology, 2011. **22**(1): p. 1-5.

19. Coifman, K., Bonanno, G, Ray, R, Gross, J, *Does repressive coping promote resilience? Affective-autonomic response discrepancy during bereavement.* J Pers Soc Psychol, 2007 **92**(4): p. 745-758.

20. Martin-Chouly, C., Morzadec, C, Bonvalet, M, Galibert, M, Fardel, O, Vernhet, L, *Inorganic arsenic alters expression of immune and stress response genes in activated primary human T lymphocytes.* Mol Immunol, 2011. **48**(6-7): p. 956-965.

21. Pert, C., *The Molecules of Emotion:The Science Behind Mind-Body Medicine.* Simon & Schuster, Inc., 1997. **New York**.

22. Nicolson, S., et al., *Comorbid Pain, Depression, and Anxiety: Multifaceted Pathology Allows for Multifaceted Treatment.* Harv Rev Psychiatry, 2009. **17**(6): p. 407-420.

23. Zheng, S., Yu, M, Lu, X, Huo, T, Ge, L, Yang, J, Wu, C, Li, F, *Urinary metabonomic study on biochemical changes in chronic unpredictable mild stress model of depression.* Clin Chim Acta, 2010. **411**(3-4): p. 204-209.

24. Béres, A., Lelovics, Z, Antal, P, Hajós, G, Gézsi, A, Czéh, A, Lantos, E, Major, T, *Does happiness help healing? Immune response of hospitalized children may change during visits of the Smiling Hospital Foundation's Artists.* Orv Hetil., 2011. **152**(43): p. 1739-1744.

25. Matsunaga, M., et al., *Association between perceived happiness levels and peripheral circulating pro-inflammatory cytokine levels in middle-aged adults in Japan.* Neuro Endocrinol Lett., 2011. **32**(4).

26. Mifsud, K., Gutièrrez-Mecinas, M, Trollope, A, Collins, A, Saunderson, E, Reul, J, *Epigenetic mechanisms in stress and adaptation.* Brain Behav Immun., 2011. **25**(7): p. 1305-1315.

27. Leonard, B., Maes, M, *Mechanistic explanations how cell-mediated immune activation, inflammation and oxidative and nitrosative stress pathways and their sequels and concomitants play a role in the pathophysiology of unipolar depression.* Neurosci Biobehav Rev, 2012. **36**(2): p. 764-85.

28. Yuen, E., Liu, W, Karatsoreos, I, Feng, J, McEwen, B, Yan, Z, *Acute stress enhances glutamatergic transmission in prefrontal cortex and facilitates working memory.* Proc Natl Acad Sci U S A, 2009. **106**(33): p. 14075-9.

29. Kendler, K., Campbell, J, *Interventionist causal models in psychiatry: repositioning the mind-body problem.* Psychol Med, 2009. **39**(6): p. 881-7.

30. Praissman, S., *Mindfulness-based stress reduction: a literature review and clinician's guide.* J Am Acad Nurse Pract., 2008 **20**(4): p. 212-6.

31. Norman, G., Hawkley, L, Luhmann, M, Ball, A, Cole, S, Berntson, G, Cacioppo, J, *Variation in the oxytocin receptor gene influences neurocardiac reactivity to social stress and HPA function: A population based study.* Horm Behav, 2012. **61**(1): p. 134-139.

32. Gamer, M., Büchel, C, *Oxytocin specifically enhances valence-dependent parasympathetic responses.* Psychoneuroendocrinology, 2012. **37**(1): p. 87-93.

33. Melville, G., Chang, D, Colagiuri, B, Marshall, P, Cheema, B, *Fifteen minutes of chair-based yoga postures or guided meditation performed in the office can elicit a relaxation response.* Evid Based Complement Alternat Med, 2012. **Epub 2012 Jan 16**.

34. Macdonald, K., Macdonald, T, *The peptide that binds: a systematic review of oxytocin and its prosocial effects in humans.* Harv Rev Psychiatry. , 2010. **18**(1): p. 1-21.

35. Macdonald, K., Feifel, D, *Dramatic Improvement in Sexual Function Induced by Intranasal Oxytocin.* J Sex Med, 2012 May;9(5):1407-10.

36. Feifel, D., Macdonald, K, Nguyen, A, Cobb, P, Warlan, H, Galangue, B, Minassian, A, Becker, O, Cooper, J, Perry, W, Lefebvre, M, Gonzales, J, Hadley, A, *Adjunctive intranasal oxytocin reduces symptoms in schizophrenia patients.* Biol Psychiatry., 2010 **68**(7): p. 678-80.

37. Clodi, M., Vila, G, Geyeregger, R, Riedl, M, Stulnig, T, Struck, J, Luger, T, Luger, A, *Oxytocin alleviates the neuroendocrine and cytokine response to bacterial endotoxin in healthy men.* Am J Physiol Endocrinol Metab., 2008. **295**(3): p. E686-91.

38. Radwany, S., VON Gruenigen, V, *Palliative and End-of-Life Care for Patients With Ovarian Cancer.* Clin Obstet Gynecol., 2012 **55**(1): p. 173-184.

39. Grant, M., Cooke, L, Williams, A, Bhatia, S, Popplewell, L, Uman, G, Forman, S, *Functional status and health-related quality of life among allogeneic transplant patients at hospital discharge: a comparison of sociodemographic, disease, and treatment characteristics.* Support Care Cancer, 2012 Nov;20(11):2697-704.

40. Matousek, R., Pruessner, J, Dobkin, P, *Changes in the cortisol awakening response (CAR) following participation in mindfulness-based stress reduction in women who completed treatment for breast cancer.* Complement Ther Clin Pract, 2011 **17**(2): p. 65-70.

CHAPTER FIFTEEN
How does Spiritual health effect psychological health and vice versa?

Spiritual health involves having meaning and direction in life. It includes the development of positive morals, ethics and values. Being spiritually healthy helps us to demonstrate love, hope, and a sense of caring for ourselves and others. It can involve organized religion, but does not necessarily have to. Spiritual health can also involve a cleric yet rarely does a cleric consult our physician regarding our overall health. Spiritual sickness is often referred to as "sin" and psychological sickness is usually thought of as mental illness. Both of these approaches make spiritual and psychological disease appear socially unacceptable. Our spiritual health greatly affects our psychological health because of our mind's connection to our soul. The sixth level chakra or celestial plane expresses our ability to love all things unconditionally. Our seventh level chakra or causal plane expresses divine wisdom in understanding the perfection in all life. By connecting the energy between the sixth and seventh chakras (unconditional love of and seeing the perfection in all things) with fifth level chakra (gaining a sense of truth and purpose) we are healing our soul.

There are many opportunities for errors in energy transmission from our soul to our mind. These errors are the cause of mental health problems. Mental illness is still very misunderstood and can be distressing for not only the patient, but friends and family of the patient. Both physiological and psychological health challenges open up the realm of the spiritual. Most people take good health for granted. When it is taken away we are given the opportunity to rediscover who we are. The loss of good health can be a gift because it provides a doorway to understanding our self in relationship to others. Expressing love and sharing quality time become very important in the lives of terminally ill people. Unfortunately we do not share this kind of love every day, not until someone gets really sick.

Chronic and incurable disease often brings with it a new sense of gratitude for life. Expression of gratitude for being alive one more day can change how someone views not only their own life but how they spend it with others. Listening to birds singing, watching flowers bloom, expressing gratitude for the sun shining, all become daily experiences that were previously taken for granted. Spiritual counseling often involves a search for meaning in terminal illnesses, confronting suffering and exploring forgiveness as well as gratitude[1].

Scientific concepts of consciousness and healing prayer

Prayer has been defined as a petition, beseeching, imploring or begging on one's behalf for an intervention from a higher power[1]. The dilemma lies in praying for someone to get well and determining if the patient's improvement is a result of the prayer, their own immune system, or mere coincidence. If prayer does work, is it potent enough to be used alone or should it be combined with orthodox therapies such as drugs or surgery? Research shows that prayer is effective in promoting the biological growth of certain forms of life under controlled laboratory conditions[2], but there are many questions with regard to prayer. Is the effect of prayer always positive or can it hurt as well as help? Are some prayer strategies better than others? Is there a best way to pray?

Prayer as a placebo

A person who knows they are being prayed for can activate healing energy in a powerful way. Prayer and thought can affect our biology[3, 4]. Studies show that what we think about our health is one of the most accurate predictors of longevity[5, 6]. But can our thoughts and prayers override our genes? Epigenetics would show it is possible. Epigenetics refer to inheritable changes beyond DNA sequence that control cell identity and morphology. Epigenetics play key roles in development and cell fate, and it highly impacts the etiology of many human diseases[7]. Outside influences such as prayer can not only affect our physiology, but our psychology as well[4, 8].

Can prayer can be harmful?

Prayer can be considered deadly if we pray that cancer or bacteria be eliminated and the bacteria and cancer cells die. Not that eliminating cancer or bacteria is a bad thing, but it is killing. If we pray negative thoughts towards bacteria and cancer, why can't prayer be just as powerful at sending negative thoughts towards a human being? If prayer works positively by itself, its beneficial power may not be due entirely to suggestion or placebo. This does not mean that placebo or suggestion effects are never involved in prayer; they can always be a factor when a person prays for himself or whether he realizes the help is associated with a pharmaceutical, a surgical procedure or something else. Many factors are involved in the outcome of prayer, including the degree of specificity or directedness of the prayer. Experiments exhibiting good scientific methodology indicate that prayer can become a balanced part of health care by significantly decreasing health problems and significantly improve quality of life[9]. The healing effects did not depend upon whether the person praying was in the presence of the organism being prayed for or praying from a distance. Healing occurred whether the healing object (fungus, bacteria, animal or human) was in a lead-lined room or a cage shielded from electromagnetic energy. It also did not matter if the person knew about the prayer or believed in prayer[10].

Scientists are discovering the exact pathways by which changes in human consciousness produce changes in human brains[11]. Our bodies respond to our thoughts and feelings with a complex array of biochemical changes that affect our organs and immune system. Emotional experiences trigger genetic changes in our cells[12]. If stress, negative thoughts and negative prayer can have an effect on our health, then it makes sense that a relaxed mind and positive prayer can affect our health as well.

How Does Prayer Work?

Because energy can organize into a coherent state, our prayers and intentions operate as highly coherent frequencies, which can change the molecular structure of our biology[13]. Highly coherent frequencies (like laser light) illuminate without losing power. If thoughts are generated as frequencies, healing intention is well-ordered light[14]. German physicist Fritz Albert-Popp constructed a photomultiplier that captures light and counts it, photon by photon. He used this device to research tiny points of light known as biophotons stored and emitted from cell DNA. The intensity of the light in organisms appears stable (ranging from a few to several hundred photons per second, per square centimeter of surface area) until it is disturbed or becomes diseased, at which point the frequency drops sharply[15]. In diseased states cancer patients emitted fewer photons. Every organism from bacteria to human beings appears to be functioning with this type of communication which can be affected by our thoughts and prayers[13]. Conscious intention is evidence that a person's thoughts and feelings create the world they live in[16, 17]. Our prayers are intentional thought processes that have the ability to affect biomolecular processes.

What do the absence of creative expression and lack of productive work have to do with the manifestation of disease?

Creativity is our soul's expression of its God self in human form. One definition of God is as a being who is a "creator". Creativity brings out the divine in us. Creativity and productivity have been shown to manifest peace and happiness[18]. People from all cultures, races and walks of life - rich, poor, educated, uneducated, famous or unknown, strive for peace and happiness. When asked what makes people happy, the one thing most studies on happiness agree is that family and relationships are the surest way to happiness[19]. Close behind are meaningful work, positive thinking, and the ability to forgive. What does not seem to make people happy are money, material possessions, intelligence, education, age, gender or attractiveness[20]. Family and friends make us happy because they make it possible for us to understand ourselves in ways we could not without these relationships. In 10 years of healing practice, I have found that the happiest people I treat are also some of the healthiest. They are people who see a higher purpose in their existence by making our world a better place. Being in service to the planet and mankind promotes health and wellness[21]. Mind-body-spirit medicine focuses on mental, emotional, social, behavioral and spiritual processes that affect health and personal growth. Yoga, meditation, prayer and expressive arts such as dance, sculpture, painting, drawing and music are often suggested in mind-body-soul practice. If both creativity and disease are our soul's way of communicating its needs, our soul's ability to heal our body comes from attaining peace and contentment through the expression of its creative and divine self. When unable to express its creativity, our soul will get bored, then depressed and ultimately diseased. When not addressed spiritual sickness spreads through our mind and into our body.

How Spiritual Health Affects Physical Health

From spiritual healing comes an opportunity for psychological and emotional growth, and ultimately behavioral and physical healing. During spiritual healing we are given the opportunity to addresses forgiveness of others and of self, gratitude for the blessings in life, as well as learning to love others and our self unconditionally. When we become physically sick we go to the physician, when we become mentally or emotionally ill we go to the psychiatrist or psychologist, and when we become spiritually distressed we go to our pastor, priest, rabbi, monk, pandit or imam. In ancient times the doctor was the holy man. In many tribal cultures the shaman is the medicine man, tribal elder, psychologist and priest[22]. Advancements in the fields of medicine, anthropology and psychology specialize in these modalities to the point where they are no longer seen to overlap. In order to understand our spirituality we must first acknowledge the existence of our soul. Our soul is our essence, which includes our consciousness, our God-self. Our soul has mass. Noetic scientists came to the conclusion that the human Soul weighs 1/3,000[th] of an ounce. East German researchers weighed more than 200 terminally ill patients just before, then immediately after their deaths. In each case the weight loss was exactly the same – 1/3,000[th] of an ounce[23]. Dr. Duncan McDougall measured the soul to be 21 grams[24]. MacDougall weighted six patients before and after death by placing their bed on an industrial sized scale sensitive to grams. The movie *21 Grams* is based on his findings. Since mass equates to energy there is a direct correlation between the concentration of energy in our biofield, chakras, body mass, thought field and the weight of our soul. Not every soul will weigh the same just as every person will not weigh the same. People who are terminally ill or dying not only lose body mass (energy) they also lose mental and spiritual energy (mass). Sick people weigh less than healthy people for many reasons.

People experiencing chronic or terminal disease often begin to embrace their spiritual side and participate more fully in life. Each precious moment becomes savored as we begin our transformation from human back to spirit. We embrace our deeper nature to discover our true identity – *we are divine*. A deep sense of purpose and more profound sense of connection with other people and our environment override superficial worries. For those who do not believe in God or a higher power, the word God can be replaced with 'joy' or 'life'. Spiritual illness is our divine self telling us it has lost connection with joy in life. The reconnection of our soul with our mental and emotional bodies will provide us with a higher quality of life. Through the reconnection of our complete self with our divine self we can find a sense of peace and serenity despite chronic or incurable illness.

References

1. Dossey, L., *Healing Words: The Power of Prayer and the Practice of Medicine*. New York: Harper Collins Publisher., 1993.
2. Harding, O., *The healing power of intercessory prayer*. West Indian Med J, 2001. **50**(4): p. 269-72.
3. Brown, C., Mory, S, Williams, R, McClymond, M, *Study of the therapeutic effects of proximal intercessory prayer (STEPP) on auditory and visual impairments in rural Mozambique*. South Med J., 2010. **103**(9): p. 864-9.
4. Vannemreddy, P., Bryan, K, Nanda, A, *Influence of prayer and prayer habits on outcome in patients with severe head injury*. Am J Hosp Palliat Care, 2009. **26**(4): p. 264-9.
5. Lovell, M., *Caring for the elderly: changing perceptions and attitudes*. J Vasc Nurs., 2006. **24**(1): p. 22-26.
6. Fry, P., Debats, D, *Cognitive beliefs and future time perspectives: predictors of mortality and longevity*. J Aging Res., 2011:367902.
7. Zachariah, R., Rastegar, M, *Linking Epigenetics to Human Disease and Rett Syndrome: The Emerging Novel and Challenging Concepts in MeCP2 Research*. Neural Plast., 2010:415825.
8. Tloczynski, J., Fritzsch, S, *Intercessory prayer in psychological well-being: using a multiple-baseline, across-subjects design*. Psychol Rep, 2002. **91**(3 Pt 1): p. 731-41.
9. Halperin, E., *Should academic medical centers conduct clinical trials of the efficacy of intercessory prayer?* Acad Med, 2001. **76**(8): p. 791-7.
10. Braud, W. and M. Schlitz, *Consciousness interactions with remote Biological systems: anomalous intentionality effects*. Subtle Energies, 1991. **2**(1): p. 1-27.
11. Kilpatrick, L., Suyenobu, B, Smith, S, Bueller, J, Goodman, T, Creswell, J, Tillisch, K, Mayer, E, Naliboff, B, *Impact of Mindfulness-Based Stress Reduction training on intrinsic brain connectivity*. Neuroimage, 2011. **56**(1): p. 290-298.
12. Pert, C., *The Molecules of Emotion:The Science Behind Mind-Body Medicine*. Simon & Schuster, Inc., 1997. **New York**.
13. McTaggart, L., *The Intention Experiment*. New York: Simon & Schuster, 2007.
14. Church, D., *The Genie in Your Genes*. Santa Rosa, CA: Elite Books, 2007.
15. Popp, F. and e. al, *Mechanism of interaction between electromagnetic fields and living organisms*. Science in China (Series C), 2000. **43**(5): p. 507-518.
16. Metzinger, T., Gallese, V, *The emergence of a shared action ontology: building blocks for a theory*. Conscious Cogn., 2003. **12**(4): p. 549-71.
17. Vandekerckhove, M., Panksepp, J, *A neurocognitive theory of higher mental emergence: from anoetic affective experiences to noetic knowledge and autonoetic awareness*. Neurosci Biobehav Rev, 2011. **35**(9): p. 2017-25.
18. Uziel, L., *Look at me, I'm happy and creative: The effect of impression management on behavior in social presence*. Pers Soc Psychol Bull., 2010. **36**(12): p. 1591-602.
19. Wallis, C., *Science of Happiness: New Research on Mood, Satisfaction*. TIME, 2005. http://www.time.com/time/magazine/article/0,9171,1015902-1,00.html.
20. http://www.usatoday.com/news/health/2002-12-08-happy-main_x.htm., *What makes people happy?* 2010.

21. Gilbert, D., *The science behind the smile. Interview by Gardiner Morse.* Harv Bus Rev., 2012. **90**(1-2): p. 84-88, 90, 152.
22. Halfax, J., *The wounder healer Shaman.* New York: The Crossroads Publishing Company, 1982. **ISBN 81-67705**.
23. Gunz, R., *1/3000th of an ounce.* Swiss Medical Journal Vita-obitus, 1993. **Weekly World News**(October 19): p. 42.
24. MacDougall, D., *Hypothesis concerning osul substance together with experimental evidence of the existance of such substances.* Am Soc of Psychical Research, 1907. **1**(5): p. 237-283.

CHAPTER SIXTEEN
How Eating Habits Affect Health

All food is made of energy due to the fact that all matter is made of energy. Different types of food have different energetic qualities. Take for instance the energy of a freshly picked organic apple and compare it to the energy of potted meat. Fresh farm eggs have a different energy than those a few weeks old. Since food is the fuel of our body, the energy in our food affects the energy in our body. Food designed by nature is the best medicine because it is fuels our body to its maximum capacity.

The biomolecular view of health and healing provides a scientific basis for orthodox medicine to prescribe pharmaceuticals such as chemotherapy and treatments such as radiation. The biofield view of healing provides a scientific foundation for many CAM treatments. Both orthodox and CAM therapies combined provide a holistic health model to heal our body of disease while decreasing or possibly eliminating the debilitating side effects that biochemical treatments cause. Both systems are dynamic, but CAM therapies involve nonlinear biological information that can alter the phase-shift of energy caused by disease. Molecular medicine involves linear pathways that take much longer to shift our body's energy toward a healing state. The energy in healthy, natural food can shift biofield energy much faster than drugs because it has a healthy resonance built-in as a baseline. Natural organic food was designed to fuel us for optimal human performance.

The human biofield is constantly moving energy to establish homeostasis. If our body becomes diseased or energetically imbalanced, the response to even a weak stimulus such as healthy food combined with biofield therapy can be dramatic. It is common knowledge that fresh natural foods provide our body with the much needed energy, vitamins and minerals that keep us active and healthy, while unhealthy foods such as processed sugars and fats deplete us of energy. Not until recently was it understood that unhealthy food can initiate the inflammatory response in our intestines creating a breeding ground for multiple diseases including fatty liver disease[1, 2]. High-carbohydrate, low-protein diets activate our inflammatory response while low-carbohydrate diets reduce inflammation. Refined sugar and other foods with high glycemic values increase insulin levels stimulating our immune response[3]. Our glycemic index measures the initial impact of a food on blood sugar levels. Increases in blood sugar trigger the release of insulin. While processed foods which contain high sugar, fat and salt, cause inflammatory responses in the body, the reverse is also true - healthy foods such as organic fruits and vegetables can initiate anti-inflammatory responses[2].

Inflammation starts with pro-inflammatory cell signaling of white blood cells to process unhealthy food particles and clear out damaged tissue. Each pro-inflammatory molecule has a complementary anti-inflammatory molecule, which neutralizes the threat and begins the healing process[4]. This very efficient system is known as an acute inflammatory response. Acute inflammation that flares up then dies down signifies a well-balanced immune system. But inflammatory responses that continue to signal the white blood cells to respond create a building up and tearing down of tissue that never ends until a drug or healthy food stops it. It is my belief that *chronic inflammation is the basis for all disease.*

Unhealthy foods contain polyunsaturated vegetable oils like safflower, sunflower, corn, and peanut which are high in linoleic acid, an omega-6 essential fatty acid, which the body converts into arachidonic acid, a harmful omega-6 fatty acid that has predominantly pro-inflammatory production and can initiate obesity[5]. These same oils contain almost no omega-3s which decrease inflammation. Anti-inflammatory Omega-3 rich oils can be found in coldwater fish such as salmon, halibut, scallops and shrimp. Flaxseeds, walnuts, sardines, soybeans, tofu and tuna are also high in omega-3 fatty acids. Omega-3 oils not only reduce inflammation, they can deter excessive clotting in our blood, help maintain fluidity of cell membranes, lower levels of lipids such as cholesterol and triglycerides, and decrease platelet aggregation (over clotting of blood). They also inhibit the thickening of arteries by decreasing endothelial cell production, increase nitric oxide (a vasodilator) to carry more oxygen rich blood to our organs, reduce the production of pro-inflammatory cytokines associated with atherosclerosis, as well as prevent cancer cell growth[6]. A deficiency in omega-3s can cause cardiovascular disease[7], type 2 diabetes[8], fatigue, an inability to concentrate, joint pain, major depression[9], dementia and Alzheimer's[10], as well as bipolar disorder[11]. There is a complex interaction between inflammatory messengers called cytokines, and prostaglandins, which are hormone-like substances that are involved in a wide range of body functions such as the contraction and relaxation of smooth muscle, the dilation and constriction of blood vessels, control of blood pressure, and modulation of inflammation[12].

Common allergens like casein and gluten (proteins found in dairy and wheat respectively) can also initiate the inflammatory response. Foods high in trans fatty acids (trans fats are formed by the addition of hydrogen to liquid fats causing them to solidify), create Low Density Lipoprotein (LDL), or "bad cholesterol," which increases the inflammatory response in arteries. Trans fats also create free radicals which can damage healthy cells and trigger inflammation[13]. The intestinal tract in particular is suspect for circulating bacteria initiating chronic inflammatory response. Intestinal bloating, frequent bouts of diarrhea or constipation, gas and pain, heartburn and acid reflux are early signs of an inflamed digestive tract[14]. These are classic symptoms of the onset of chronic physiological disease.

Ayurvedic medicine suggests the two main areas affecting our health and wellness are the foods we eat and the lifestyle we live. By changing our eating habits and lifestyle, we greatly influence our physical, mental and emotional health. Foods based on the five elements of earth, water, fire, air and ether can be instrumental in balancing energetic disharmony and overriding disease[15]. Due to the homeostatic nature of our biofield, as one aspect of our body, mind and/or spirit, becomes unbalanced, at least one other if not all the other aspects become unstable as well. The onset of elemental physical imbalances is first initiated by spiritual imbalance. For instance the earth chakra is red. The earth chakra represents our relationship with money, family and survival. It functions to protect us physically. Without food, clothing, shelter and familial support our health is compromised. When our earth chakra is out of balance our soul is unsettled. When this occurs our mental and emotional stability becomes unstable whereby we can get so angry we "see red". We are also said to be "green with envy" when feeling jealous because our heart chakra is shut down. This chakra aligns with the air element. Jealousy is the inability to want only the best for everyone and everything. We are described as being "blue" when we are sad and our throat chakra is contracted due to grief or loss. Many times we have a "lump" in our throat due to an energy blockage. Blue is the color of the ether element. Cowards are described as "yellow bellies" which correlates to the fire element and the solar plexus or third chakra. Their power is weakened and are said to have no "guts".

One powerful way to rebalance these biofield elements is by making healthy food choices. Elemental imbalances are caused by either contraction or expansion of one or more of the five elements. If you have a contracted earth element, a high concentration of fruit in your diet can exacerbate contracted earth issues because they contain a lot of sugar which is associated with earth – thus earth becomes more contracted. If the earth element is expanded, eat earth foods (red) to contain it down and bring it back to balance (see Table 1). You will know the earth element is balanced if you feel safe, have regular bowel movements, and appropriate boundaries. If earth is contracted eat air foods (green) to break it up. Constipation, knee or neck pain[16], boundary issues such as interfering in other people's business or becoming too reclusive are all signs of contracted earth. If the water element is contracted, eat water foods to expand it. Water foods are orange in color. There is an imbalance of water element associated with disorders or diseases of the feet, breast or ovary/testes. Emotional water imbalances express as constant crying or wanting to cry, addiction or attachment issues, and unwarranted overreaction to situations. If water is expanded eat fire foods to calm it. If the element of fire is contracted then eat fire foods to expand it. Our fire is contracted if we have ulcers, boils, acne, rashes, or feelings of self loathing. Contracted fire also causes liver, stomach and gall bladder disease. Contracted fire is often released in explosive emotion. In order to release pent up fire exercise (especially swim), find a secluded area and yell really loud, or chop wood. All these exercises help to release fire in an appropriate manner. Fire is expanded if we have headaches, rage management issues or inflammatory eye problems. If fire is expanded eat water foods to cool it down. If air is expanded eat fire foods to burn it up. Fire thrives on air. If air is contracted, eat fire and earth foods to calm it. Jitteriness, constant chatter, and judging others are signs of expanded air. Shoulder, kidney or ankle disorders are all signs of contracted air. Depression, unrelenting grief, deafness and joint pain are all signs the ether element is contracted. Expanded ether presents as spaciness, fog brain, dementia or Alzheimer's disease. There are certain foods associated with each specific element and fortunately they can be categorized most often by their color and their growing location (Table 1).

Element	Earth	Water	Fire	Air	Ether
Color	Red	Orange	Yellow	Green	Blue/Purple
Where they are grown	Underground	Surface of earth up to two feet	From two to six feet above ground	Above six feet	Everywhere
Counter balance	Eat air foods to expand and earth foods to contract	Eat water foods to expand and fire foods to contract	Eat fire and air foods to expand and water and earth foods to contract	Eat air foods to expand and earth foods to contract	Eat earth foods to contract and ether foods to expand
Types	beets watermelon cranberries strawberries potatoes red meat raw sugar	oranges melons cauliflower broccoli carrots cucumber brown rice chicken and eggs	peppers onions garlic all grains: rye wheat barley oats corn	apples nuts lemons limes lettuces apricots peaches pears	blueberries blackberries blue potatoes sprouts edible flowers blue corn yogurt

Table 1. Elemental foods with associated color and location

There are several foods which have multiple elements. Bananas for instance are earth (they can constipate you), air (they grow in trees), fire (they are yellow) and water (because they are the ovary of the banana tree). Bananas are very popular and eaten by many cultures of people. Fruit that grows in trees are air, but their color can add another element to their affect in our body. Sweet red apples have earth element in them whereas green apples do not. Keep in mind that fruits contain sugar so earth is always going to be a part of that food source.

Earth Foods – mostly red in color, these foods grow in and directly above the earth (see figure1). Earth foods include red beets, watermelon, leeks, turnips, cranberries, tomatoes, parsnips and potatoes. Red meats and sugar cane are also earth foods. Many earth vegetables are rich in lycopene, which can reduce the risk of prostate cancer[17]. Lycopene is an antioxidant compound that gives tomatoes and other fruits and vegetables their color. Foods rich in lycopene include tomatoes and tomato products, watermelon, guava, papaya and rose hips. Lycopene has been associated with lower risk of heart disease, macular degenerative disease, prostate cancer and lipid oxidation (the damage to normal fat molecules that can cause inflammation and disease)[18]. It has also been reported to lower LDL ("bad" cholesterol), enhance the immune system, protecting our DNA[19]. A major claim for lycopene is in the prevention and treatment of cancers of the lung, stomach, bladder, cervix, skin, and especially, prostate. Lycopene is a powerful antioxidant that blocks the action of free radicals, which are activated oxygen molecules that can damage cells[20, 21]. Other beneficial earth foods include those with resveratrol, such as red wine, red and purple grapes, blackberries, cranberries, mulberries and even peanuts. Resveratrol contains polyphenol which neutralizes free radicals and inhibits inflammation[22]. Earth foods balance fear and phobias, relieve stubbornness, allow for flexibility regarding boundaries, and promote ease of movement in the neck and knees. When earth is contracted we not only feel neck and low back pain, as well as inflammation in our knees, but also constipation, anxiousness and fear; we get stuck in old patterns and initiate the growth of tumors and fibroids. When earth is expanded, we often feel paranoid and worry, suffer from irritable bowels and colon problems, have inappropriate boundaries, feel defensive, insecure and become lazy. Chronic and incurable diseases of the earth element include osteoarthritis, ruptured vertebrae, irritable bowel syndrome (IBS), Crohn's Disease, colon, blood and bone cancers.

Water Foods – mainly orange in color, these foods grow in and on top the ground up to 2 feet (see figure 1). They include squashes, papaya, lentils, persimmons, carrots, yams, tangerines, cantaloupes, and cauliflower. They also include chicken, fish, and rice. Water foods are a good source of flavonoids which are polyphenolic compounds that are found in plant-derived foods and beverages such as oil, tea, and red wine. The biological activities of flavonoids cover a broad spectrum, from anticancer[23] and antibacterial activities[24] to the modulation of the inflammatory response[25]. Flavonoids have also been shown to modulate neuroglial interactions during central nervous system development which opens the possibility for their use as therapeutic strategies for neurodegenerative diseases[26]. Powerful flavonoids stave off inflammation and blood vessel damage caused by poor diets. Water foods are also high in alpha-carotene. These include sweet potatoes, carrots, winter squash, and cantaloupe. Alpha-carotene converts to vitamin A in the body boosting our immune system. It has been identified as an anti-aging compound[27]. Water foods are also a good source of beta-cryptoxanthin, found in papaya and tangerines. This carotenoid is important to bone and cell growth[28] as well as vision[29] Turmeric or curcumin is a good spice to balance water. The antioxidant properties of curcumin can counteract the body's negative responses to high-fat food as well as reduce inflammation[30]. It is considered to be of therapeutic value in the treatment and prevention of obesity-related cancers[31]. Water foods regulate feelings of possessiveness, emotional trauma, attachment, edema, lymph flow, spleen, and bladder problems. Contracted water is responsible for addictions, possessiveness and attachment. It can also cause the development of excess mucus, hip and foot pain, intense menstrual cramps as well as lymphedema. Expanded water initiates compulsive behavior, endometriosis, swelling, PMS symptoms, prostate trouble and feelings of being overwhelmed. Chronic and incurable water diseases include bladder and kidney disease, chronic edema, breast cancer and lymphoma.

Fire Foods – mostly yellow in color, these foods grow between 2 feet and 6 feet above ground (see Table 1). Fire foods are a good source of limonoids, whose benefits include lower cholesterol and protection against breast, skin and stomach cancers[32]. Fire foods also contain bromelain found in pineapple which has a cardio-protective effect. Bromelain also ameliorates rejection-induced arterial wall remodeling, prevents thrombin-induced human platelet aggregation, and also reduces thrombus formation[33]. Bromelain has been shown to increase the death rate of breast cancer cells[34]. Foods such as grains, squash, yellow bell peppers and hot peppers, as well as lentils, are examples of fire foods. Corn contains lutein, which is a xanthophyll, and one of 600 known naturally occurring carotenoids. Lutein has shown beneficial effects in the prevention of vascular diseases[35]. Fire foods are also rich in capsaicin, which is a colorless compound responsible for the spicy taste of hot chili peppers, cayenne peppers and jalapenos. Capsaicin is used both internally and as a topical application for pain management[36]. It has been shown to prevent the replication of prostate cancer cells[37]. Emotional fire imbalances include issues ranging from anger management problems to cowardice. Contracted fire causes acid reflux, gallbladder and liver disease, blindness, headaches, lack of clarity and seething emotions. Expanded fire exacerbates ego-power trips, blaming others, explosive rage, loud voice, inflammation, boils and infections, eye redness and indigestion problems. Chronic and incurable fire disease includes migraine headache, blindness, all eye diseases, chronic indigestion, as well as liver, stomach and pancreatic cancer.

Air Foods – mainly green in color these foods grow 6 feet and above the ground (see figure 1). Air foods are a good source of chlorophyll which can decrease the risk of liver and prostate cancer. Chlorophyll can be found in watercress, leeks, arugula, and parsley. Other green air foods which help purge the body of potential carcinogens are kale, Brussels sprouts and broccoli[38]. Air foods are also a good source of catechins found in green tea. Catechins have been reported to be beneficial in inhibiting invasion and metastasis of prostate cancer[39], therapeutic in non-alcoholic fatty liver disease[40], and have a promising role in the reversal of age-related loss of neuronal plasticity and recovery after neuronal lesions associated with aging[41]. Air foods also supply apigenin and luteolin, flavonoids found in celery and parsley. Research shows apigenin and luteolin can have neuroprotective/disease-modifying properties in various neurodegenerative disorders, including Alzheimer's disease[42]. Air imbalances include jittery uncontrolled movement in the body and thought patterns. Contracted air presents as shoulder pain, shallow breathing, and also lack of compassion and judgment. Expanded air produces dry itchy skin, impatience, scattered thinking, rapid talking and weak ankles. Chronic and incurable air diseases include panic and anxiety attacks, heart disease, Chronic Obstructive Pulmonary Disease (COPD), emphysema, lung and esophageal cancer.

Ether foods – usually blue or purple in color, these include eggplant, purple cauliflower and cabbage, purple grapes, sprouts, edible flowers, blue corn, blueberries black berries and fermented foods such as yogurts and yeast. Red cabbage, eggplant, grapes, and berries in particular are a good source of antioxidants which improve cognitive brain function[43]. They can also reduce the risk of cancer, stroke, heart disease and Alzheimer's disease[43]. Ether foods are a good source of indoles. These can be found in purple cauliflower and purple cabbage. Indoles are derived from sulfur compounds in cruciferous vegetables and have been found to slow the metabolic rate of carcinogens[44]. Ether foods such as berries have a phytochemical known as ellagic acid that can lessen the effect of estrogen in promoting breast cancer cell growth[45]. Ether foods should be eaten if the ether element is contracted. Contracted ether is expressed as grief, hopelessness, low self esteem, close mindedness, and joint pain. Earth foods should be eaten if ether is expanded. Expanded ether presents as a lack of self expression, pride, chaotic life, spaciness and separation from Source/God. In particular expanded ether is responsible for severe depression, dementia and Alzheimer's disease. Other chronic or incurable diseases involving the ether element are arthritis, tinnitus, deafness, asthma, and brain tumors.

Eating healthy food not only changes the way energy is regulated throughout our biofield, but provides much needed energy and anti-aging benefits for a long and healthy life. You will know you are on a path to health and healing when you crave healthy foods because "like attracts like". You will know you are on an unhealthy path when you crave unhealthy foods. Try to eat foods from each color/element group every day. Remember to move your body to keep energy flowing throughout your field. If you are currently facing the challenge of a chronic or incurable disease, a change in diet and lifestyle will enhance your quality of life by increasing your energy level and improving mental clarity.

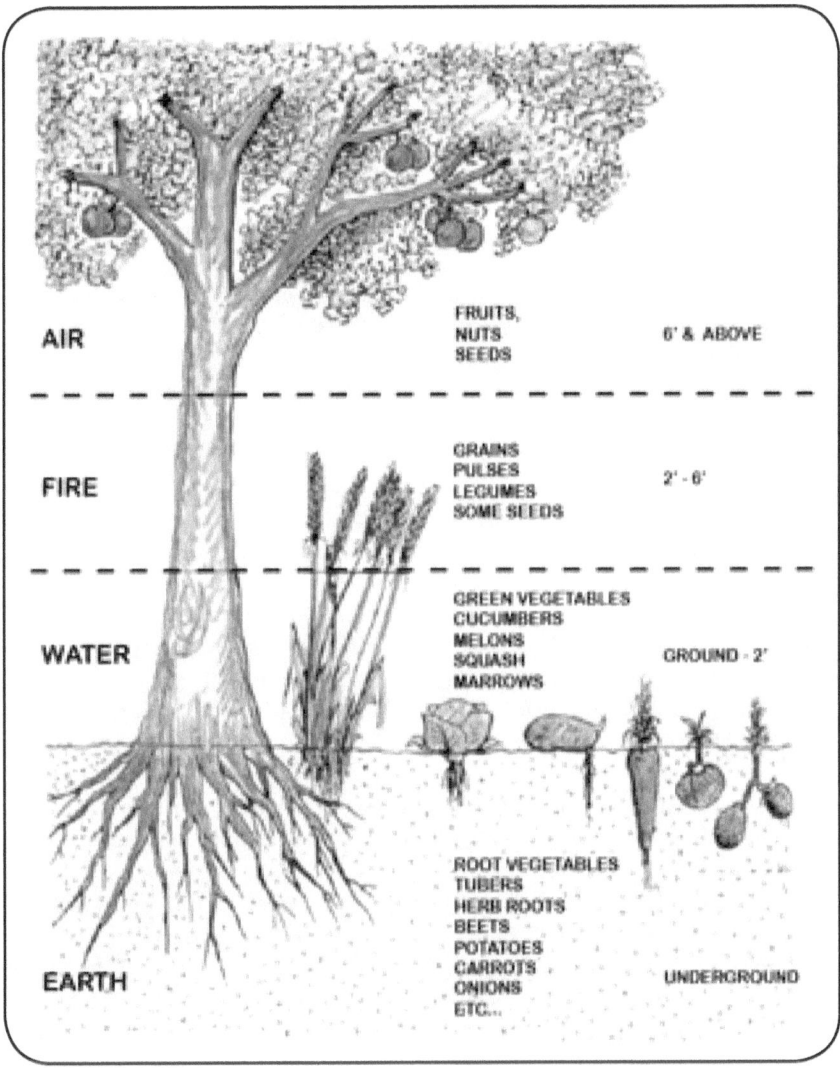

Figure 1. Location of different food groups

References

1. Zelber-Sagi, S., Ratziu, V, Oren, R, *Nutrition and physical activity in NAFLD: an overview of the epidemiological evidence*. World J Gastroenterol, 2011. **17**(29): p. 3377-89.

2. Calder, P., Ahluwalia, N, Brouns, F, Buetler, T, Clement, K, Cunningham, K, et al, *Dietary factors and low-grade inflammation in relation to overweight and obesity*. Br J Nutr., 2011. **106** (Suppl 3): p. S5-78.

3. Venkatesan, R., Devi, A, Parasuraman, S, Sriram, S, *Role of community pharmacists in improving knowledge and glycemic control of type 2 diabetes*. Perspect Clin Res, 2012 **3**(1): p. 26-31.

4. Cuneo, A., Autieri, M, *Expression and function of anti-inflammatory interleukins: the other side of the vascular response to injury*. Curr Vasc Pharmacol, 2009. **7**(3): p. 267-276.

5. Alvheim, A., Malde, M, Osei-Hyiaman, D, Hong, Lin, Y, Pawlosky, R, Madsen, L, Kristiansen, K, Frøyland, L, Hibbeln, J, *Dietary linoleic acid elevates endogenous 2-AG and anandamide and induces obesity*. Obesity, 2012 Oct;20(10):1984-94.

6. Calder, P., *Fatty acids and inflammation: the cutting edge between food and pharma*. Eur J Pharmacol, 2011. **668**(Suppl 1): p. S50-58.

7. Lee, J., Jarreau, T, Prasad, A, Lavie, C, O'Keefe, J, Ventura, H, *Nutritional assessment in heart failure patients*. Congest Heart Fail, 2011. **17**(4): p. 199-203.

8. Kouchak, A., Djalali, M, Eshraghian, M, Saedisomeolia, A, Djazayery, A, Hajianfar, H, *The effect of Omega-3 fatty acids on serum paraoxonase activity, vitamins A, E, and C in type 2 diabetic patients*. J Res Med Sci, 2011. **16**(7): p. 878-884.

9. Nahas, R., Sheikh, O, *Complementary and alternative medicine for the treatment of major depressive disorder*. Can Fam Physician, 2011 **57**(6): p. 659-663.

10. Hashimoto, M., Hossain, S, *Neuroprotective and ameliorative actions of polyunsaturated fatty acids against neuronal diseases: beneficial effect of docosahexaenoic acid on cognitive decline in Alzheimer's disease*. J Pharmacol Sci, 2011. **116**(2): p. 150-162.

11. Bazan, N., Molina, M, Gordon, W, *Docosahexaenoic acid signalolipidomics in nutrition: significance in aging, neuroinflammation, macular degeneration, Alzheimer's, and other neurodegenerative diseases*. Annu Rev Nutr, 2011. **31**(321-351).

12. Westover, A., Hooper, S, Wallace, M, Moss, T, *Prostaglandins mediate the fetal pulmonary response to intrauterine inflammation*. Am J Physiol Lung Cell Mol Physiol, 2012 Apr 1;302(7):L664-78.

13. Iwata, N., Pham, M, Rizzo, N, Cheng, A, Maloney, E, Kim, F, *Trans fatty acids induce vascular inflammation and reduce vascular nitric oxide production in endothelial cells*. PLoS One, 2011. **6**(12): p. e29600.

14. Pendyala, S., Walker, J, Holt, P, *A High-Fat Diet is Associated with Endotoxemia that Originates from the Gut*. Gastroenterology, 2012 May;142(5):1100-110.

15. Aggarwal, B., Prasad, S, Reuter S, Kannappan, R, Yadev, V, Park B, Kim, J, Gupta, S, Phromnoi, K, Sundaram, C, Prasad, S, Chaturvedi, M, Sung, B, *Identification of novel anti-inflammatory agents from Ayurvedic medicine for prevention of chronic diseases: "reverse pharmacology" and "bedside to bench" approach*. Curr Drug Targets Inflamm Allergy, 2011. **12**(11): p. 1595-653.

16. Sills, F., *The Polarity Principle: energy as a healing art*. North Atlantic Books, 2002. **Berkley, CA**.
17. Yang, C., Lu, Y, Chen, H, Hu, M, *Lycopene and the LXRa agonist T0901317 synergistically inhibit the proliferation of androgen-independent prostate cancer cells via the PPARγ-LXRa-ABCA1 pathway*. J Nutr Biochem, 2012 Sep;23(9):1155-62.
18. Ozten-Kandaş, N., Bosland, M, *Chemoprevention of prostate cancer: Natural compounds, antiandrogens, and antioxidants - In vivo evidence*. J Carcinog, 2011. **10**: p. 27.
19. Lim, U., Song, M, *Dietary and lifestyle factors of DNA methylation*. Methods Mol Biol, 2012. **863**(359-376).
20. Khoo, H., Prasad, K, Kong, K, Jiang, Y, Ismail, A, *Carotenoids and their isomers: color pigments in fruits and vegetables*. Molecules., 2011. **16**(2): p. 1710-1738.
21. Palozza, P., Parrone, N, Catalano, A, Simone, R, *Tomato lycopene and inflammatory cascade: basic interactions and clinical implications*. Curr Med Chem, 2010. **17**(23): p. 2547-2563.
22. Wang, G., Guo, X, Chen, H, Lin, T, Xu, Y, Chen, Q, Liu, J, Zeng, J, Zhang, X, Yao, X, *A resveratrol analog, phoyunbene B, induces G2/M cell cycle arrest and apoptosis in HepG2 liver cancer cells*. Bioorg Med Chem Lett., 2012 Mar 1;22(5):2114-8.
23. Park, E., Pezzuto, J, *Flavonoids In Cancer Prevention*. Anticancer Agents Med Chem, 2012 Oct 1;12(8):836-51.
24. Santos, R., Kushima, H, Rodrigues, C, Sannomiya, M, Rocha, L, Bauab, T, Tamashiro, J, Vilegas, W, Hiruma-Lima, C, *Byrsonima intermedia A. Juss.: Gastric and duodenal anti-ulcer, antimicrobial and antidiarrheal effects in experimental rodent models*. J Ethnopharmacol., 2012 Mar 27;140(2):203-12.
25. Zampini, I., Villena, J, Salva, S, Herrera, M, Isla, M, Alvarez, S, *Potentiality of standardized extract and isolated flavonoids from Zuccagnia punctata for the treatment of respiratory infections by Streptococcus pneumoniae: In vitro and in vivo studies*. J Ethnopharmacol, 2012 Mar 27;140(2):287-92.
26. Nones, J., Spohr, T, Gomes, F, *Effects of the flavonoid hesperidin in cerebral cortical progenitors in vitro: indirect action through astrocytes*. Int J Dev Neurosci, 2012 Jun;30(4):303-13.
27. Ferrucci, L., Perry, J, Mattein,i A, Perola, M, Tanaka, T, Silander, K, Rice, N, Frayling, M, *Common variation in the beta-carotene 15,15'-monooxygenase 1 gene affects circulating levels of carotenoids: a genome-wide association study*. Am J Hum Genet., 2009 **84**(2): p. 123-133.
28. Uchiyama, S., Yamaguchi, M, *beta-cryptoxanthin stimulates cell differentiation and mineralization in osteoblastic MC3T3-E1 cells*. J Cell Biochem., 2005. **95**(6): p. 1224-1234.
29. Tan, J., Wang, J, Flood, V, Rochtchina, E, Smith, W, Mitchell, P, *Dietary antioxidants and the long-term incidence of age-related macular degeneration: the Blue Mountains Eye Study*. Ophthalmology, 2008. **115**(2): p. 334-341.
30. Kim, K., Lee, E, Park, J, Lee, J, Kim, J, Choi, H, Kim, B, Lee, H, Lee, K, Yoon, S, *Curcumin Attenuates TNF-a-induced Expression of Intercellular Adhesion Molecule-1, Vascular Cell Adhesion Molecule-1 and Proinflammatory Cytokines in Human Endometriotic Stromal Cells*. Phytother Res. , 2012 Jul;26(7):1037-47.

31. Shehzad, A., Khan, S, Sup, Lee, Y, *Curcumin molecular targets in obesity and obesity-related cancers*. Future Oncol, 2012. **8**(2): p. 179-190.

32. Manners, G., *Citrus limonoids: analysis, bioactivity, and biomedical prospects*. J Agric Food Chem, 2007. **55**(21): p. 8285-9824.

33. Ley, C., Tsiami, A, Ni, Q, Robinson, N, *A review of the use of bromelain in cardiovascular diseases*. Zhong Xi Yi Jie He Xue Bao., 2011. **9**(7): p. 702-710.

34. Dhandayuthapani, S., Perez, H, Paroulek, A, Chinnakkannu, P, Kandalam, U, Jaffe, M, Rathinavelu, A, *Bromelain-Induced Apoptosis in GI-101A Breast Cancer Cells*. J Med Food, 2011 Apr;15(4):344-9.

35. Lo, H., Tsai, Y, Du, W, Tsou, C, Wu, W, *A Naturally Occurring Carotenoid, Lutein, Reduces PDGF and H2O2 Signaling and Compromised Migration in Cultured Vascular Smooth Muscle Cells*. J Biomed Sci, 2012 **19**(1): p. 18.

36. Anand, P., Bley, K, *Topical capsaicin for pain management: therapeutic potential and mechanisms of action of the new high-concentration capsaicin 8% patch*. Br J Anaesth, 2011 **107**(4): p. 490-502.

37. Mori, A., Lehmann, S, O'Kelly, J, Kumagai, T, Desmond, J, Pervan, M, McBride, W, Kizaki, M, Koeffler, H, *Capsaicin, a component of red peppers, inhibits the growth of androgen-independent, p53 mutant prostate cancer cells*. Cancer Res, 2006. **66**(6): p. 3222-3229.

38. Zhang, Y., Talalay, P, Cho, C, et al., *A major inducer of anticarcinogenic protective enzymes from broccoli: isolation and elucidation of structure*. Proc Natl Acad Sci U S A, 1992. **89**: p. 2399–2403.

39. Connors, S., Chornokur, G, Kumar, N, *New insights into the mechanisms of green tea catechins in the chemoprevention of prostate cancer*. Nutr Cancer, 2012. **64**(4-22).

40. Masterjohn, C., Bruno, R, *Therapeutic potential of green tea in nonalcoholic fatty liver disease*. Nutr Rev, 2012. **70**(1): p. 41-56.

41. Andrade, J., Assunção, M., *Protective Effects of Chronic Green Tea Consumption on Age-related Neurodegeneration*. Curr Pharm Des, 2012. **18**(1): p. 4-14.

42. Rezai-Zadeh, K., Ehrhart, J, Bai, Y, Sanberg, P, Bickford, P, Tan, J, Shytle, R, *Apigenin and luteolin modulate microglial activation via inhibition of STAT1-induced CD40 expression*. J Neuroinflammation, 2008. **5**: p. 41.

43. Zhao, Y., Zhao, B, *Natural antioxidants in prevention and management of Alzheimer's disease*. Front Biosci (Elite Ed), 2012. **1**(4): p. 794-808.

44. Adler, S., Rashid, G, Klein, A, *Indole-3-carbinol inhibits telomerase activity and gene expression in prostate cancer cell lines*. Anticancer Res, 2011. **31**(11): p. 3733-3777.

45. Wang, N., Wang, Z, Mo, S, Loo, T, Wang, D, Luo, H, Yang, D, Chen, Y, Shen, J, Chen, J *Ellagic acid, a phenolic compound, exerts anti-angiogenesis effects via VEGFR-2 signaling pathway in breast cancer*. Breast Cancer Res Treat, 2012 Aug;134(3):943-55.

CHAPTER SEVENTEEN
The use of both Conventional and CAM modalities to heal

In order to become more involved in our own disease prevention and healing process, we must educate ourselves to the options available as alternatives or complements to conventional medicine. In the past 10 years there has been a distinct trend toward the integration of complementary and alternative medicine (CAM) therapies with the practice of conventional medicine. According to a recent report from the American Hospital Association (AHA), more hospitals are now offering complementary and alternative medicine (CAM) in addition to traditional medical care. A 2011 AHA report shows hospitals across the US are responding to patient demand by integrating complementary and alternative medicine (CAM) services with conventional medical services currently available[1].

The results of a survey released by Health Forum, [a subsidiary of the American Hospital Association (AHA)] and Samueli Institute, [a non-profit research organization that investigates healing oriented practices], showed more than 42 percent of the 714 responding hospitals indicated they offer one or more CAM therapies, (up from 37 percent in 2007)[1]. Health maintenance organizations (HMOs) are covering CAM therapies, and a growing number of physicians use CAM therapies in their practices as insurance coverage for these therapies is increasing. Integrative medicine centers and clinics are being established, many with close ties to medical schools and teaching hospitals[2].

In a survey conducted at Kaiser Permanente on the use of CAM therapies among clinicians of the Kaiser Permanente Northern California Region, nearly 90 percent of clinicians reported recommending at least one type of alternative therapy in the previous 12 months and 2/3 of physicians and 3/4 of ob/gyn clinicians were interested in recommending alternative therapies[3]. Message therapy, chiropractics and acupuncture, along with vitamin and mineral supplements were the highest recommended CAM therapies.

In 2007 americans spent $33.9 billion out-of-pocket (uninsured) on complementary and alternative medicine (CAM) over the previous 12 months[4]. Approximately 38 percent of adults use some form of CAM for health and wellness or to treat a variety of diseases and conditions. According to data from the 2007 National Health Interview Survey (NHIS), the most recent year for available statistics. Almost 4 out of 10 adults had used CAM therapy in the past 12 months, with the most commonly used therapies being non-vitamin, non-mineral, natural products (17.7 percent) and deep breathing exercises (12.7 percent)[5]. Results from the NHIS found that approximately one in nine children (11.8 percent) used CAM therapy in the past 12 months (3.9 percent) and chiropractic or osteopathic manipulation (2.8 percent). For both adults and children in 2007, when concerns about cost delayed the delivery of conventional care, they were more likely to use CAM than when the cost of conventional care was not a concern. Between 2002 and 2007 an increased use in acupuncture, deep breathing exercises, massage therapy, meditation, naturopathy, and yoga was seen among adults. CAM use for head or chest colds showed a decrease from 2002 to 2007 (9.5 percent to 2.0 percent). As of July 30, 2009 Americans spent almost a third as much money out-of pocket on herbal supplements and other alternative medicines as they did on prescription drugs[1]. Out-of-pocket spending on herbal supplements, chiropractic visits, meditation, and other forms of complementary and alternative medicines (CAM) was estimated at $34 billion in a single year. Overall out-of-pocket expenditures for CAM treatments accounted for 1.5 percent of the $2.2 trillion spent on health care during the year prior to the survey. Excluding intercessory prayer, statistics show chiropractic manipulation, herbal medicine, massage, and homeopathy were the therapies most commonly used by the general population[6]. More users were women, middle aged, and more educated. The ailments most often associated with CAM use include back pain, depression, insomnia, severe headache or migraine, and stomach or intestinal illnesses. People who see traditional health care providers more frequently were also the most likely to use complementary care services[7]. Insurance coverage associated with CAM therapy use found that for people who sought the services of practitioners performing physical manipulation (e.g., chiropractors and massage therapists), full insurance coverage, partial insurance coverage, and use of the therapy for wellness were all associated with the high-frequency use of such providers. Sixty-three percent of the people using CAM services having full insurance coverage made eight or more visits to CAM practitioners annually. Only 17 percent of those reporting no insurance coverage made eight or more visits.

How Can Conventional Western Medicine Benefit from Complementary and Alternative Medicine (CAM) and vice versa?

CAM cannot exist without conventional and orthodox therapies to draw from; and conventional medicine is not sustainable if it does not focus on patient-oriented clinics and treatment options with fewer side effects. In contrast to conventional medicine, CAM does not diagnose or fix, it uses our body's innate ability to heal itself, along with the principle that consciousness shapes our understanding of health and disease[8]. CAM cannot exist apart from our conscious awareness and intention. Today's medicine must also integrate and appreciate multiple social and cultural approaches to healing, and that appreciation must be as important as scientific information and technology. The most effective treatments appear to be orthodox medicine working together with CAM including changes in diet and lifestyle [9, 10].

A questionnaire was distributed to 1757 registered physicians, nurses and physiotherapists in surgical wards at seven university hospitals from spring 2010 to spring 2011. The questionnaire included classification of 21 therapies in conventional, complementary, alternative and integrative medicine, and whether patients were recommended these therapies. Questions concerning knowledge, research, and patient communication about CAM were also included. A total of 737 (42.0 percent) questionnaires were returned. Therapies classified as complementary were massage, manual therapies, yoga and acupuncture. Alternative therapies were herbal medicine, dietary supplements, homeopathy and healing. Classification of integrative therapy was low, and unfamiliar therapies were Bowen therapy, iridology and Rosen method. Therapies recommended by >40 percent of the participants were massage and acupuncture. Their knowledge of and research about CAM was valued as minor or none at all by 95.7 percent and 99.2 percent respectively. Importance of possessing knowledge about it was valued as important by 80.9 percent. It was believed by 61.2 percent that more research funding should be addressed to CAM research, 72.8 percent were interested in reading CAM-research results, and 27.8 percent would consider taking part in such research. More than half of the participants (55.8 percent) were open to learning more about CAM therapies and CAM research. Communication about CAM between patients and the health care professionals was found to be rare[11]. Lack of understanding and research knowledge contributed to much of the misinformation regarding CAM therapies. Not only do health care practitioners need to educate themselves, but patients must bring information to their providers to help them understand patient desires and needs.

There is a lack of equality granted to CAM insurance coverage as compared with conventional medical coverage. This is unfortunate for all health care practitioners. Physicians who practice conventional medicine are experiencing economic pressures to reduce the amount of time they spend with patients, contributing to burnout among medical staff and endangering their patients iatrogenically. Politicians are getting involved as the public is calling for more affordable health care. A new paradigm must be embraced in order to address all aspects of this dilemma. It is clear that science and technology have resulted in a vastly improved understanding, diagnosis, and treatment of disease, but the emphasis on science and technology to the exclusion of human interaction has also served to limit the development of a model that humanizes health care[12]. The healing of a patient must include more than the biology and chemistry of their physical body; by necessity, it must include the mental, emotional and spiritual aspects. Western medicine treats mental and emotional disorders with psychotherapy and psychiatry, which are separate modalities from physiological medicine which treats the body. It is to our benefit that psychotherapy has become more socially acceptable, but there are still many who think only "crazy" people seek help from a psychiatrist and only "sinful" people seek spiritual counseling.

Untreated Acute Illness Leads to Chronic Disease

Physiological disease first appears as acute illness. Acute illness is a condition that will eventually resolve itself over time or with help from pharmaceuticals, natural remedies or self healing. Such illnesses include fever, cold, coughs, congestion, throat infections, stomach or abdominal problems, rashes or skin injuries, muscle and joint problems, bladder or vaginal infections, respiratory infections, headaches and eye infections. Because we are unaware that acute illness is information, we attempt to quiet it with analgesics, calcium carbonate, ice packs, heat packs, antibiotics, cough syrup and cold remedies. But the message is still there - our mind, emotions and soul are energetically unstable, therefore without addressing the emotional, psychological and spiritual aspects of acute illness, chronic disease will set in. Chronic diseases are defined as those that persist indefinitely. They cannot be prevented by vaccines and cannot be cured by medication. Such prolonged illnesses lead to ongoing pain, suffering, disability and diminished quality of life. Chronic diseases are diseases of long duration and generally slow progression[13]. Chronic diseases, such as heart disease, diabetes, stroke, cancer, as well as chronic respiratory diseases such as chronic obstructive pulmonary disease (COPD) and emphysema, are by far the leading cause of death and disability in the United States, accounting for about 70 percent of all deaths[14] and about 75 percent of the $2.6 trillion the nation spends on health care services[15]. Out of the 36 million people who died from chronic disease in 2008, 29 percent were under 60 and half were women. Eighty percent of these diseases occur in low and middle income patients. According to the World Health Organization[16], chronic diseases are now the major cause of death and disability worldwide. Non-communicable conditions, including cardiovascular diseases (CVD), diabetes, obesity, cancer and respiratory diseases, now account for 59 percent of the 57 million US deaths annually and 46 percent of the global disease. High cholesterol, high blood pressure, obesity, smoking and alcohol cause the majority of the chronic disease problems. This reflects a significant change in dietary habits, physical activity levels, and tobacco use worldwide as a result of industrialization, urbanization, economic development and the globalization of the food market[17]. In order to remedy or at least decelerate the affects of chronic disease we must redefine disease and how to treat it physically, mentally, emotionally, culturally, and spiritually. Below are listed the leading causes of death along with comorbidity and suggested CAM therapies and lifestyle changes.

Disease and the Mind-Body Connection – dealing with Comorbidity

Cardiovascular Disease (CVD)

Heart attacks and strokes kill about 12 million people every year; another 3.9 million die from hypertension and other heart conditions. More than one billion adults worldwide are overweight; at least 300 million of them are clinically obese. About 75 percent of CVD can be attributed to the majority risks: high cholesterol, high blood pressure, low fruit and vegetable intake, inactive lifestyle and tobacco use. Sustained behavioral interventions have been shown to be effective in reducing population risk factors. According to the Centers for Disease Control (2011), almost one in four Americans has some form of cardiovascular disease[18]. It has been the main cause of death in the US every year since 1900, except for 1918, the year of the Spanish flu pandemic. People with Coronary Artery Disease (CAD), or the accumulation of hardened atherosclerotic plaques or deposits along the inner lining of the arteries, are candidates for chest pain (angina) and heart attack (myocardial infarction). CAD has three major clinical syndromes: angina pectoris, myocardial infarction and sudden cardiac death. CAD can also present as silent ischemia, which is blockage of blood flow to the heart without pain or discomfort. Ischemic cardiomyopathy is a condition in which a portion of the heart muscle no longer functions due to its not having sufficient blood flow. The onset and progress of cardiovascular disease has a significant body of research indicating that it is directly affected by psychological factors such as stress, depression and social isolation[19]. Behavior factors regarding diet, exercise, weight control and adherence to prescribed medications also play a role. Main areas of focus for the psychological issues associated with CVD are health promoting behavioral changes and adherence to recommended medical treatment. Clinical depression may be found in up to 44 percent of cardiac patients over the course of their disease. Depression can increase blood pressure, alter blood clotting, affect heart rhythm and raise cholesterol and insulin levels[20]. Coping factors include concerns regarding any medical or surgical intervention initiating patients' concerns about future ability to function, earn a living, perform family activities, participate in exercise, as well as deal with financial stresses of rising medical costs. A regimen of yoga, meditation and relaxed breathing exercises would be feasible at any level and provide some of the doctor recommended behavioral changes needed for stress reduction.

From the Ayurvedic and Traditional Chinese Medicine (TCM) diagnostics, heart disease can be associated with a lack of energy flow into and out of the heart chakra. Heart disease can also manifest as a lack of unconditional love of others and of self. Without the free flow of unconditional love through our heart chakra, blocked energy can squeeze the life out of us. Biofield therapies such as Polarity Therapy, Reiki, Therapeutic Touch, Trager and Healing Touch assist in regulating erratic electrical impulses in the heart known as arrhythmias[21]. Biofield practitioners often work in conjunction with physicians and other health care professionals to achieve lowered blood pressure, decreased stress and anxiety, as well as offset the side effects of treatments such as chemotherapy and radiation[22]. Ayurveda addresses depression as a result of the patient's separation from Source, God, or the Universal Life Force. Therapy involves focusing on progressive relaxation and abdominal breathing which can relieve fear and anxiety about heart disease diagnosis and prognosis. Opportunities for growth include embracing the joy in life instead having it squeezed out when the patient is overworked and unfulfilled.

Cancer

Cancer is the uncontrolled growth and spread of abnormal cells due to errors in cell communication. These cells undergo a series of alterations due to genetic or environmental causes, which result in their inability to respond normally to signals that control cell growth, differentiation and cell death. Tumors thrive on oxygen and other nutrients easily found in blood supplies and are therefore able to induce blood vessel growth (angiogenesis), in and around the tumor, through the secretion of growth factors[23]. The causes of cancer include environmental factors (i.e., tobacco use, radiation, chemicals), immunoendocrine factors (i.e., infectious organisms, viruses, hormones), and genetic factors (inherited mutations, metabolism-related mutations). The most prevalent cancers include breast, prostate, lung, and colorectal cancer. There are seven common warning signs and symptoms of cancer, including: a change in bowel habits or bladder function; sores that do not heal; unusual bleeding or discharge; a lump or thickening in any part of the body (e.g. breast); indigestion or trouble swallowing; a change in a mole or wart; or a cough or hoarseness that does not go away. Cancer treatment involves surgery, chemotherapy, radiation or hormonal therapy and also biological therapy (aka immunotherapy or biological response modifiers) which function to enhance the ability of our immune system to fight cancer and infections[24]. Comorbid psychopathology with cancer patients includes depression, post traumatic stress disorder, personality disorders, chemo or fog brain (problems with attention and concentration, verbal and visual memory and word finding difficulty along with motor skill deficits) due to the effects of chemotherapy. Quality of Life (QOL) outcomes are improved by evaluating and developing psychosocial and behavioral interventions; focusing on the education of medical information about cancer and its treatments; and also providing patients and their support members with effective coping mechanisms for stress and emotional responses that accompany cancer diagnosis[25]. When cancer is diagnosed, it typically affects the entire family system. For siblings of children with cancer, feelings of loss, loneliness, anger, jealousy due to an imbalance of attention, and guilt are frequently experienced, although gains in personal growth, maturity and independence are also seen.

An Ayurvedic assessment of cancer patients addresses the symptoms of fever, fatigue, pain or skin changes. Fever is associated with the fire element. Fatigue and burnout are a lack of fire. Fire is the element of spirit, courage, will and desire. Yoga, meditation, prayer and expressive arts such as dance, painting, sculpture and music are suggested therapies for cancer patients. Cancer patients' opportunity for spiritual growth include the search for meaning in life by confronting suffering, expressing our creativity through the arts, and pursuing higher quality relationships with those we love. The urgency of cancer requires that we act quickly and dramatically to make life changes.

Diabetes

Diabetes occurs when an individual develops a dysfunction in insulin production and/or insulin action, which induces an inability to metabolize blood glucose. Insulin is a hormone produced by beta cells within the pancreas. Insulin is necessary for cellular metabolic function of glucose to be carried within our blood stream. Type I Diabetes (formerly known as Juvenile diabetes or insulin-dependent diabetes) occurs when the body destroys the insulin-producing beta cells of the pancreas in what appears to be an autoimmune reaction. Without insulin production an individual will die if they do not receive insulin from another source. Since insulin is destroyed by the digestive process it cannot be taken orally and must be delivered by some form of injection. Type II Diabetes (previously called adult-onset diabetes) begins with the development of insulin resistance. When cells do not properly use available insulin, the demand for insulin production increases and the pancreas eventually fails to produce sufficient insulin. Risk factors for Type II diabetes include obesity, aging, family history of diabetes and physical inactivity.

As of January 2011 it is estimated that 18.8 million people in the US have diabetes (approx. 8.3 percent of the population) and 79 million have pre-diabetes[26]. About 90 – 95 percent of all people with diabetes have Type II diabetes leaving 5 to 10 percent with Type I. Symptoms include increased thirst and urination, fatigue, increased hunger, and weight loss. Diabetes of either type increases risk for retinopathy and blindness; peripheral neuropathy, numbness and pain in the feet and legs, stroke, myocardial infarction, peripheral vascular problems, end stage renal disease, and impaired healing, particularly in the limbs and amputation of lower extremities[27]. High blood glucose (BG) levels that induce tremor, sweating, restlessness, nausea, headaches, nervousness, intense hunger, palpitations, increased respiration rate, weakness or tingling sensation around the mouth. Low BG levels induce confusion, dysnomia (word finding confusion), emotional instability, impaired thinking and judgment, dizziness, amnesia, stupor, difficulty walking or coordinating motor movement, and at extreme levels, delirious or psychotic-like functioning, unconsciousness, or seizure. Self management and adherence to regimen is essential to the QOL controls of diabetic patients. Insulin dosage and timing of food intake are important in preventing diabetic complications. Comorbid psychopathology includes depression, anxiety, and dysregulated eating. Incorporating diaphragmatic breathing and a vivid recreation of a calming meditation can result in relieving these problems[28]. Behavioral therapists can assist patients with insulin regimen and dietary control[29]. Energy medicine assessment includes thought patterns that evoke a loss of "sweetness" for life. Multiple disappointments can lead to cynical thoughts of "what's the use?" "Everyone else gets all the treats, when is it going to be my turn?" What evolves is a deep sorrow for what "might have been" and a need to control what is left in life. Energy field theory associates diabetes with the earth element releasing stubborn behavior and giving up control over expectations, having more flexible boundaries, and releasing the need for events to unfold in a certain way. Opportunities for growth include the realignment of old thinking patterns towards gratitude for what one has (no matter how small) which can shift the patient's energy towards healing rather than regret. Understanding no one is too old to pursue their passion in life can lead to transformations from sorrow to joy and a love of life even into advanced age. *The purpose of life is to pursue our deepest desire.*

Stroke

Stroke is a leading cause of death in the United States. Over 800,000 people die in the U.S. each year from stroke[30]. A stroke occurs when a clot blocks the blood supply to the brain or when a blood vessel in the brain bursts. Stroke can cause death or significant disability, such as paralysis, speech difficulties, and emotional problems[31]. About 85 percent of all strokes are ischemic, where blood flow to the brain is blocked by blood clots or fatty deposits called plaque in blood vessel linings[32]. A hemorrhagic stroke occurs when a blood vessel bursts in the brain. Blood accumulates and compresses the surrounding brain tissue. Symptoms of stroke include paralysis and weakness on one side of the body, problems with thinking, awareness, attention, learning, judgment, memory, understanding or forming speech. Other signs of a stroke include difficulty controlling or expressing emotions, numbness or strange sensations, pain in the hands and feet that worsens with movement and temperature changes, and depression[33].

Energy medicine assesses stroke as giving up on or rejecting life. The person would rather die than change. This is a form of stubbornness that is counterproductive to our soul and depletes our QOL. Stroke victims have difficulty adapting to change and accepting all aspects of life, past present and future[34]. Opportunities for growth include the realignment of old thinking patterns towards flexibility in what life has to offer, and accepting change by viewing it as an adventure instead of a curse. Renewed health involves letting go of past regrets and understanding our past and present experiences form who we have become today.

Chronic Respiratory Disease

Chronic respiratory disease is characterized as a disease state with limited airflow and which is not fully reversible. It is associated with an abnormal inflammatory response of the lungs to noxious particles or gases. Airflow limitation leads to symptoms of breathlessness and fatigue. Because airflow limitations are not fully reversible, the chronic baseline deficit may increase but will not return to normal physiology even if the patient no longer has symptoms. Chronic respiratory disease is progressive meaning that the baseline will change negatively over time.

Common symptoms of chronic respiratory disease such as asthma, emphysema, chronic bronchitis, and Chronic Obstructive Pulmonary Disease (COPD) include cough, sputum (mucus) production, and dyspnea (a feeling of breathlessness or shortness of breath). A history of exposure to risk factors for COPD very commonly includes smoking, but also includes repetitive exposure to toxic substances such as paint or industrial chemicals. With COPD inflammation of the airways leads to mucus hyper-secretion and remodeling of airways, causing obstruction and narrowing of the air passages. Oxidative stress destroys lung tissue and leads to the development of emphysema. Vascular changes contribute to the development of pulmonary hypertension increasing pressure in the vessels that carry deoxygenated blood from the right side of the heart to the lungs for oxygenation. Prolonged increases can lead to enlargement and failure of the right side of the heart. Approximately 24 million adults have impaired lung function in the US. It is the fourth leading cause of death behind myocardial infarction, cancer and stroke. Eight million annual office visits, 1.5 million emergency room visits, 726,000 hospitalizations and 119,000 deaths are attributed to COPD[35]. Although death rates for COPD have declined among U.S. men between 1999 (57.0 per 100,000) and 2006 (46.4 per 100,000) in the United States, there has been no significant change among death rates in women (35.3 per 100,000 in 1999 and 34.2 per 100,000 in 2006). The economic burden is substantial, with an annual cost that exceeds $29.5 billion[36]. Cigarette smoking accounts for the majority of causes with only 15 percent of COPD as work related (asbestos abatement, chemical inhalation, etc.). Once the diagnosis of chronic respiratory disease is made, communicating the seriousness of the condition is essential to promoting optimal patient and family coping. Disease progression includes mechanical ventilators for patients with severe exacerbations that could accompany intubations. Further lung damage results from chronic episodes[24]. Comorbidities include bronchial cancer, tuberculosis, sleep apnea and left heart failure. Cigarette smoking is a major risk factor in the development of COPD and accounts for the vast majority of patients with the illness, so smoking cessation is an essential part of the treatment.

Self-efficacy refers to an individual's belief regarding his or her ability to perform a particular behavior or set of behaviors[37]. People who feel they lack control view their health as primarily influenced by outside factors such as chance, enabling friends, or the environment, tend to minimize the importance of behavioral change such as smoking cessation. Many people with COPD tend to become overly focused on physical symptoms, leading to increased levels of anxiety and depression[38]. Less anxiety and depression have been found in patients with higher levels of social support[39].

An energy medicine assessment of chronic respiratory disease involves an air imbalance related to the brachial plexus area of the body. A desire to smoke would be brought on by dyspnea (shortness of breath) and the need to breathe. Many people who have difficulty breathing or do not consciously inhale deeply are attracted to smoking because smoking involves the act of inhaling deeply. Many smokers were exposed to second hand smoke as children[40]. Typical symptoms include a rigid diaphragm, shallow breathing, shoulder pain, a feeling of breathlessness, depression, lack of compassion, and a fear of "taking life in". Energy medicine treatment would include adding more water and healthy oils to the patient's diet as well as brachial plexus work which includes the ribs and shoulders. Energy follows the breath. Having a high functioning lung and diaphragm capacity is the catalyst for energy flow throughout the brachial plexus area and our entire body. Air is the most quickly distributed element in the body and unlike food which takes minutes to hours to digest and change our body chemistry, each inhalation of air enters our bloodstream immediately. Aside from maintaining basic life functions, the breath is one of our most powerful tools for transformation: for burning up toxins, releasing stored emotions, changing body structure and changing our consciousness. Conscious breathing helps us cultivate inner stillness and presence. It also provides an intimate pathway to introspection allowing us to be present without judgment or analysis. Opportunities for growth include learning to breathe more effectively and self empowerment through the projection of safe boundaries.

HIV / AIDS

HIV is a retrovirus that binds to cells that carry the CD4 molecule, which include T-helper lymphocytes (T-cells), monocytes, and macrophages, which are primary components of the immune system. The function of the T-cell is to respond to attacks from foreign agents by mounting an immune response, which should neutralize the foreign agent. HIV takes over T-cells and multiplies, destroying more T-cells and ultimately damaging the body's defenses against infection. As the number of T-cells drops, people with HIV become more susceptible to other infections and certain types of cancer, which their bodies are no longer able to fight. As their immune system becomes increasingly deficient, they are diagnosed with AIDS (acquired immunodeficiency syndrome).

Worldwide, more than 34 million people were living with human immunodeficiency virus (HIV) in 2010, including 38 million adults and 2.3 million children. There were 2.7 million new HIV infections and 1.8 million people died from AIDS in 2010. During 2010 some 2.7 million people became newly infected with the virus, including an estimated 390,000 children. Despite a significant decline in the estimated number of AIDS-related deaths over the last five years, there were still an estimated 1.8 million AIDS-related deaths in 2010. Of those, 20 percent had undiagnosed HIV infections[41]. The CDC estimates approximately 40,000 Americans become infected with HIV/AIDS each year. HIV is transmitted via three primary routes: sexual, blood-borne and perinatal (through childbirth). Because the disease is spread through modifiable behaviors, understanding both transmission rates and patterns, and the course of the illness progression is imperative for prevention. Abstinence programs along with comprehensive sexuality education helps with prevention; although abstinence only programs generally do not work[42].

Morbidity and mortality related to HIV has decreased significantly as a result of highly active antiretroviral therapy (HAART). Antiretroviral drugs can have severe side effects such as causing liver problems, diabetes, abnormal fat distribution, high cholesterol, increased bleeding in patients with hemophilia, decreased bone density, skin rash, pancreatitis and nerve problems[43]. Stressful life conditions necessitate adherence to drug therapies and behavioral changes. Relationship instability, housing instability, history of incarceration and unemployment all increase the likelihood of non-adherence. Studies show that even perceived stress is enough to reduce adherence especially when co-morbid with depression[44]. Depression, anxiety, antisocial behavior, death anxiety, changed life expectations; borderline personality disorders are often symptomatic of HIV/AIDS diagnoses[45]. High social support has been found to decrease depression in HIV/AIDS patients. Coping efforts include utilizing both guided imagery and diaphragmatic breathing to reduce stress levels related to the disease prognosis[24].

HIV/AIDS Ayurvedic assessment includes an imbalance of the water element which encompasses relationship issues, feelings of isolation, inability or unwillingness to nurture oneself, and feelings of sexual guilt. This manifests in the form of contracted water which progresses towards emotional instability. Water is a powerful carrier, mediator and producer of energy. It has the ability to transform and capture both physical elements and subtle energy forces. Opportunities for growth include learning to love self and others unconditionally. Accepting differences in all people, including all races, cultures and belief systems is important to balancing the water element in our biofield. It is important to feel powerful and capable despite feeling "different" in society. *All people have a right to be here* whether the reasons why are clear to others or not.

Obesity

According to the American Medical Association, a person must have a body mass index (BMI) of 25 to 29.9 in order to be considered overweight. Obesity is divided into mild (BMI of 30 to 34.9), moderate (BMI of 35 to 39.9) and severe (BMI above 40). Center for Disease Control Statistics, state that during the past 20 years, there has been a dramatic increase in obesity in the United States and rates remain high. More than one-third of U.S. adults (35.7%) and approximately 17% (or 12.5 million) of children and adolescents aged 2—19 years are obese[46]. Obesity is the result of a complex interaction of genetic, behavioral and environmental influences, indicating that a multidisciplinary approach to treatment is necessary[47]. Obesity poses serious health risks, decreased quality of life and increased health care expenditures. While obesity has reached epidemic proportions, treatment options have expanded. These include psychological treatments, pharmaceuticals and bariatric procedures (stomach stapling and LAP-BAND surgery). Cognitive behavioral techniques focus on how thoughts influence attitude and behaviors about eating and exercise. A common thinking pattern that interferes with weight loss is an "all-or-nothing" thinking about food as either "bad" or "good" and that one needs to exercise for long periods of time in order for it to be helpful. Addressing body image is one of the most importance issues in changing lifestyle and behavioral patterns regarding food and exercise. When the clinician is a behavioral health provider, it is important that he or she consult with a supervising physician or that the patient is under the care of a medical provider during the course of their treatment. The use of multiple treatment modalities to maximize outcomes requires the behavioral health clinician to become familiar with treatment approaches used by professionals in other disciplines (e.g., bariatric surgeons, registered dieticians, exercise physiologists).

Energy medicine assessment of obesity addresses body fat as a protective armor. The patient needs protection from external threats (teasing, bullying, abuse, etc.) and from the internal feelings of insecurity and low self esteem. Fat is a way of separating and protecting us from the world. Fat build-up is due to anger towards environmental situations such as poverty, lack of opportunity, abuse, family and work relationships, or lack of nutritional education. Other life events that impact weight include marriage and relationship problems, job loss, injury, family illness and death of a loved one. Opportunities for growth include learning to love and appreciate the uniqueness of all individuals especially self. Loving our soul instead of the temporary, superficial housing of our soul (our physical body) allows acceptance to come natural.

By addressing the leading causes of death through spiritual, psychological, emotional and physiological pathways, the health care industry predicts an estimated 80 percent of heart attacks can be prevented or delayed through lifestyle changes[48]; more than 50 percent of cancers could be prevented through tobacco cessation and healthier diet and lifestyle[49]; chronic obstructive pulmonary disease, (10 million diagnosed, but perhaps 24 million affected), could result in a higher QOL[50] and type II diabetes, the leading cause of non-traumatic amputations and blindness in persons 20 – 64 and leading cause of end stage renal failure, could be prevented or at least reversed[51]. Hypertension (high blood pressure) and hyperlipidemia (high cholesterol) are also preventable causes of death in the US. Hypertension is the leading cause of congestive heart failure, hemorrhagic stroke, ischemic coronary disease and cerebrovascular disease[52].

Chronic diseases can attribute to or exacerbate fatigue, hopelessness, lack of motivation and depression in a patient, which may also impact a person's consideration of the future. Depression is associated with higher rates of smoking and lower levels of physical activity[53]. Depression also increases the risk of obesity, hypertension, heart disease and stroke[54]. The side effects of pharmaceuticals alone are a significant reason to begin integrating CAM practices and practitioners with behavioral health modification and therapists, primary care physicians, and psychologists who are trained to address many of the aforementioned diseases and their comorbidites. The combination of energy medicine practitioners and spiritual counselors with the above mentioned panel of health care providers will provide the most affective care available. Cancer Treatment Centers of America (CTCA) are a good example of how integrative medicine improves QOL and empowers patients. A survey of 358 patients who participated in the Patient Empowered Care pilot was conducted from October 5, 2009 through January 7, 2010. More than 90% of CTCA patients feel they received the information they needed to understand their medical condition and treatment options. CTCA has integrative cancer treatments that expand the boundaries of conventional care using traditional methods for fighting cancer, such as surgery, radiation, chemotherapy, and immunotherapy, with supportive therapies, that include nutritional support, naturopathic medicine, mind-body medicine, oncology rehabilitation, pain management, and spiritual support. Statistical results of this type of care can be seen on their website www.cancercenter.com. This type of treatment could be applied to all types of chronic diseases.

Health care providers cannot assume people either want to change or that their health is a prime motivation for change. Making decisions about preventative care, alternative medicine or behavioral changes is influenced by the patient's social and cultural background, cognitive and emotional health, as well as family support system[55]. Unlike most conventional interventions, such as expensive testing and pharmaceuticals, energy psychology and energy medicine interventions, tailored to the chronic disease population and delivered in primary care settings, are inexpensive and have demonstrated significant outcomes. In order to slow the soaring health care expenditures of chronic disease and reduce disparities, integrating medicine should be seen as a major solution to our health care problem.

References

1. Fenwick, M., Hutcheson, D, *More Hospitals Offering Complementary and Alternative Medicine Services.* http://www.aha.org/presscenter/pressrel/2011/110907-pr-camsurvey.pdf, 2011.

2. Lundgren, J., Ugalde, V, *The demographics and economics of complementary alternative medicine.* Phys Med Rehabil Clin N Am, 2004. **15**: p. 955–961.

3. Gordon, N., Sobel, D, *Use of and Interest in Complementary and Alternative Therapies Among Clinicians and Adult Members of the Kaiser Permanente Northern California Region: Results of a 1996 Survey.* The Permanente Journal, 1999. **3**(2): p. 44-55.

4. Briggs, J., *Out-of-Pocket CAM Use in 2007.* National Center for Complementary and Alternative Medicine (NCCAM) 2009. http://nccam.nih.gov/news/camstats/2007(Retrieved February 23, 2012).

5. Barnes, P., Bloom, B, Nahin R, *Complementary and Alternative Medicine Use Among Adults and Children: United States. National health statistics reports; no 12.* Hyattsville, MD: National Center for Health Statistics. 2008, 2007.

6. Public., I.o.M.U.C.o.t.U.o.C.a.A.M.b.t.A., *Complementary and Alternative Medicine in the United States.* Washington (DC): National Academies Press (US), 2005. **ISBN-10: 0-309-09270-**.

7. Wolsko, P., Eisenberg, D, Davis, R, Ettner, S, Phillips, R, *Insurance coverage, medical conditions, and visits to alternative medicine providers: Results of a national survey.* Intern Med., 2002. **162**(3): p. 281–287.

8. Wilber, K., *Consciousness and Healing: Integral Approaches to Mind-Body Healing.* St Louis (MO): Elsevier/Churchill/Livingstone, 2005.

9. Ausfeld-Hafter, B., Hoffmann, S, Seibold, F, Quattropani, C, Heer, P, Straumann *Status of alternative medicine in Crohn disease and ulcerative colitis patents: a questionnaire survey.* Forsch Komplementarmed Klass Naturheilkd, 2005. **12**(3): p. 134-138.

10. Quimby, E., *The use of herbal therapies in pediatric oncology patients: treating symptoms of cancer and side effects of standard therapies.* J Pediatr Oncol Nurs, 2007. **24**(1): p. 35-40.

11. Bjersa, K., Stener, V, Fagevik ,O, *Knowledge about complementary, alternative and integrative medicine (CAM) among registered health care providers in Swedish surgical care: a national survey among university hospitals.* BMC Complement Altern Med, 2012. **12**(1): p. 42.

12. Schlitz, M., Tina Amorok, T, Micozzi, M, *Consciousness and Healing: Integral Approaches to Mind-Body Medicine.* Churchill Livingston/Elsevier, 2005. **St Louis, MO**.

13. Organization, W.H., *Chronic diseases and health promotion.* http://www.who.int/chp/en/, 2006. **Retrieved January 10, 2012**.

14. Kung, H., Hoyert, D, Xu, J, Murphy, S, *Deaths: final data for 2005. National Vital Statistics Reports* Department of Health and Human Services, 2008. **56**(10).

15. Martin, A., Lassman, D, Whittle, L, Catlin, A, *Recession contributes to slowest annual rate of increase inhealth spending in five decades.* Health Aff (Millwood), 2011. **1**: p. 11-22.

16. Organization, W.H., *Chronic diseases and health promotion.* . http://www.who.int/chp/en/, 2011. **Retrieved January 10, 2011**.

17. McMichael, A., *Integrating nutrition with ecology: balancing the health of humans and biosphere.* Public Health Nutr, 2005. **8**(6A): p. 706-15.

18. Prevention, C.f.D.C.a., *Cardiovascular Disease Facts Sheet.* 2012 Retrieved April 20. http://www.cdc.gov/omhd/AMH/factsheets/cardio.htm.

19. Albus, C., *Psychological and social factors in coronary heart disease.* Ann Med, 2010. **42**(7): p. 487-94.

20. Todini, L., Majorana, M, Luciani, A, Orso, L, *Depression and cardiovascular risk. relationship, pathogenetic mechanisms, treatment and patient management. A review of the literature.* Clin Ter., 2012. **163**(2): p. e77-e84.

21. Anderson, J., Taylor, A, *Biofield therapies in cardiovascular disease management: a brief review.* Holist Nurs Pract., 2011. **25**(4): p. 199-204.

22. Hart, L., Freel, M, Haylock, P, Lutgendorf, S, *The use of healing touch in integrative oncology.* Clin J Oncol Nurs., 2011. **15**(5): p. 519-25.

23. Axnick, J., Lammert, E, *Vascular lumen formation.* Curr Opin Hematol, 2012. **19**(3): p. 192-198.

24. Boyer, B., Paharia, M, *Comprehensive Handbook of Clinical Health Psychology.* Hoboken, NJ: John Wiley & Sons, Inc, 2008.

25. van der Mei, S., Dijkers, M, Heerkens, Y, *Participation as an outcome measure in psychosocial oncology: content of cancer-specific health-related quality of life instruments.* Qual Life Res, 2011. **20**(10): p. 1617-27.

26. Association, A.D., *Diabetes Basics Data from the 2011 National Diabetes Fact Sheet (released Jan. 26, 2011).* www.diabetes.org/diabetes-basics/diabetes-statistics, 2011. **Retrieved May 4, 2012**.

27. Vaidyanathan, J., Choe, S, Sahajwalla, C, *Type 2 diabetes in pediatrics and adults: Thoughts from a clinical pharmacology perspective.* J Pharm Sci, 2012 May;101(5):1659-71.

28. Kulur, A., Haleagrahara, N, Adhikary, P, Jeganathan, P, *Effect of diaphragmatic breathing on heart rate variability in ischemic heart disease with diabetes.* Arq Bras Cardiol, 2009. **92**(6): p. 423-9, 440-7, 457-63.

29. Najafian, B., Alpers, C, Fogo, A, *Pathology of human diabetic nephropathy.* Contrib Nephrol, 2011. **170**:: p. 36-47.

30. Miniño, A., Murphy, S, Xu, J, Kochanek, K *Deaths: Final data for 2008 [PDF-2.9M]. National Vital Statistics Reports.* National Center for Health Statistics: Hyattsville, MD, 2011. **59**(10).

31. Muqtadar, H., Testai, F, *Single Gene Disorders Associated With Stroke: A Review and Update on Treatment Options.* Curr Treat Options Cardiovasc Med, 2012 Jun;14(3):288-97.

32. Lloyd-Jones, D., Adams, R, Carnethon M, et al, *Heart disease and stroke statistics—2009 update. A report from the American Heart Association Statistics Committee and Stroke Statistics Subcommittee.* Circulation 2009. **119**((suppl I)): p. e21-e181.

33. Ostwald, S., Wasserman, J, Davis, S, *Medications, Comorbidities, and Medical Complications in Stroke Survivors: The CAReS Study.* Rehabil Nurs., 2006. **31**(1): p. 10–14.

34. Hay, L., *Heal Your Body.* Hay House: Carlsbad, CA, 1976.

35. Disease, C.O.P., *COPD data 2006*. COPD International, 2011. http://www.copd-international.com/library/statistics.htm.
36. Dalal, A., Liu, F, Riedel, A, *Cost trends among commercially insured and Medicare Advantage-insured patients with chronic obstructive pulmonary disease: 2006 through 2009*. Int J Chron Obstruct Pulmon Dis, 2011. **6**: p. 533-42.
37. Lichtenstein, E., and Glasgow, R, *Smoking cessation: What have we learned over the past decade?* Journal of Consulting and Clinical Psychology, 1992. **60**(4): p. 518-527.
38. Jiang, X., He, G, *Effects of an uncertainty management intervention on uncertainty, anxiety, depression, and quality of life of chronic obstructive pulmonary disease outpatients*. Res Nurs Health, 2012 Aug;35(4):409-18.
39. Spandler, H., Stickley T, *No hope without compassion: the importance of compassion in recovery-focused mental health services*. J Ment Health, 2011. **20**(6): p. 555-66.
40. Man Ping Wang, M.P., Sai Yin Ho, Tai Hing Lam, *Parental Smoking, Exposure to Secondhand Smoke at Home, and Smoking Initiation Among Young Children*. Nicotine and Tobacco Research, 2011(doi: 10.1093/ntr/ntr083).
41. Control, C.f.D., *HIV Surveillance --- United States, 1981--2008*. MMWR, 2011. **60**(21): p. 689-693.
42. Regassa, N., Kedir, S, *Attitudes and practices on HIV preventions among students of higher education institutions in Ethiopia: the case of Addis Ababa University*. East Afr J Public Health., 2011. **8**(2): p. 141-54.
43. Paydary, K., Emamzadeh-Fard, S, Khorshid, H, Kamali, K, Seyed, Alinaghi, S, Mohraz, M, *Safety and Efficacy of Setarud (IMOD TM) Among People Living with HIV/AIDS: A Review*. Recent Pat Antiinfect Drug Discov, 2012. **7**(1): p. 66-72.
44. Amirkhanian, Y., Kelly, J, Kuznetsova, A, DiFranceisco, W, Musatov, V, Pirogov, D, *People with HIV in HAART-era Russia: transmission risk behavior prevalence, antiretroviral medication-taking, and psychosocial distress*. AIDS Behav, 2011. **15**(4): p. 767-77.
45. Kong, M., Nahata, M, Lacombe, V, Seiber, E, Balkrishnan, R, *Association Between Race, Depression, and Antiretroviral Therapy Adherence in a Low-Income Population with HIV Infection*. J Gen Intern Med, 2012 Sep;27(9):1159-64.
46. Prevention, C.f.D.C.a., *Obesity in the US*. Adult Obesity Facts, 2010. http://www.cdc.gov/obesity/data/adult.html.
47. Obesity, *Centers for Disease Control and Prevention*. http://www.cdc.gov/obesity/, 2012. **retrieved March 17, 2012**.
48. Hankinson, S., Colditz, G., Manson,. J, Speizer, F., (Eds), *Healthy Women, healthy lives: A guide to preventing disease, from the landmark Nurses' Health Study*. New York: Simon & Schuster, 2001.
49. Kostopoulou, V., Katsouyanni, K, *The truth-telling issue and changes in lifestyle in patients with cancer*. J Med Ethics, 2006. **32**(12): p. 693-7.
50. Preventions, C.f.D.C.a., *COPD*. http://www.cdc.gov/copd/, 2010. **retrieved March 15, 2012**.
51. Prevention, C.f.D.C.a., *Diabetes*. http://www.cdc.gov/diabetes/, 2012. **retrieved March 12, 2012**.
52. Health, N.I.o., *The sixth report of the Joint National committee on prevention, detection,*

evaluation and treatment of high blood pressure. Archives of Internal Medicine, 1997. **157**: p. 2413 – 2446.

53. Anda, R., Williamson, D., Escobedo, L., Mast, E., Giovino, G., & Remington, P, *Depression and the dynamics of smoking: A national perspective.* Journal of the American Medical Association, 1990. **264**: p. 1541 – 1545.

54. Davidson, K., Jonas, B., Dixon K., & Markovitz, J, *Do depression symptoms predict early hypertension incidence in young adults from the CARDIA study?* Archives of Internal Medicine, 2000. **160**: p. 1495 – 1500.

55. Bandura, A., *Self-efficacy: The exercise of control.* New York: W.H. , 1999.

CHAPTER EIGHTEEN
How Do I know which CAM is best for my condition?

When chronic or terminal diseases are diagnosed, depending on the severity, our minds go into shock. Especially with a diagnosis like cancer which can be life threatening creating confusion on how to proceed. Unfortunately in most cases we must be our own advocate. Patients with life threatening illnesses want to know what their options are, which ones work best, and if they are safe. More people are turning to Complementary and Alternative Medicine (CAM) because of its holistic approach that includes not just the physiological, but mental, emotional and spiritual aspects of disease. Being diagnosed with a life threatening illness is very emotionally and psychologically disruptive; people look to CAM to find solace despite challenge and to avoid or counteract the side effects of conventional therapies.

It is self empowering to be involved in our own healing process. It is important to work with a physician who will include you in the decision making process of your health care plan. If you are afraid chemotherapy will kill you before the cancer does – tell your doctor. If your preference is to use CAM as adjunct therapy for the side effects of chemotherapy or radiation - tell your doctor. Feeling comfortable that your clinician will work with you instead of talking down to you is a first step in the healing process. If you do not feel comfortable sharing your interest in CAM therapies with your physician due to concerns of dismissal or fear of ridicule, find another doctor. Diagnosis can be confusing and scary. Information is power - taking an active part in your own healing process will empower you as a patient and as a person.

Internist and integrative medical doctor, Larry Dossey suggests the popularity of CAM therapies is largely due to the fact they help people find meaning in their lives when they need it most. No matter how technologically advanced modern medicine may be, if it does not honor the place of meaning in illness, it loses the allegiance of those it serves[1]. There are many situations in which surgery, pharmaceuticals and radiation can play an active role in the healing process, but many times they are the *only* options given patients. The side effects of pharmaceuticals and radiation can often be reduced or at times completely eliminated with complementary medicine. Alternative approaches such as change in diet and lifestyle (smoking cessation and more exercise), along with weekly or monthly treatments can leave pharmaceuticals as a last option instead of the first. Believing "the doctor always knows best" has rendered patients both powerless and uninformed in their own healing process. Working with a physician to decide which treatment is best educates both the physician and the patient.

In the past patients received information about diagnosis, prognosis and treatment primarily from physicians. Patients from past generations generally did not challenge physicians' advice or question prescribed treatments, because they were passive recipients of information that was filtered and dispensed by health care providers. Since the advent of the internet patients have been interactive in decision making due to broader access to information, increased patient autonomy, and patients' ability to acquire the tools to research clinical conditions and interpret medical data[2]. A different relationship has emerged from this new model of decision making in which the physician recognizes the need of the patient to participate as a partner in interpreting medical information and selecting treatment options[3].

The Institute of Medicine (IOM) has targeted patient-centeredness as an important area of quality improvement. A major aspect of patient-centeredness is respect for patient's values, preferences, and expressed needs[4]. Protocols for gaining this understanding and translating it to quality care in the clinical setting are lacking. Patients not only need to understand technical outcomes, but participate in decision-making, treatment options, be given emotional support; and be provided with a comfortable patient-doctor relationship. Physicians who leave a patient to wait for up to 2 hours to get 15 minutes of the doctor's time are not running patient centered clinics. Physician centered clinics do not initiate healing environments. CAM practitioners set aside 1 to 1.5 hours of their time to gather information about their patient, to understand what is presenting emotionally, psychologically and spiritually, as well as physically. CAM practitioners do *not* diagnose pathology or psychoanalyze, they assess for imbalances in the patient's biofield. Body pulses (not blood pressure or heart rate), voice tones, body position, and emotional expression all play a part in the assessment of the patient. Often treatment feedback forms are supplied to the patient for assurance of a satisfactory visit. As a Polarity Practitioner I listen first, assess second, treat third then ask for feedback. Involvement in one's own healing process immediately activates healing, whether from the thought that the patient will get well, the actual treatment occurring, or both. To determine which CAM is best it is necessary to research different modalities. There are so many to choose from it can be difficult to know where to begin. The following CAM treatments are modalities I have tried and received beneficial results. *This list is in no way meant to be a comprehensive in-depth description of each modality.* It is simply an overview of a few CAM options available. There are dozens, if not hundreds, more. It is up to you, the reader, to do the research and provide the groundwork for which modality is the best for you personally. I have both given and received hundreds of polarity therapy sessions; therefore, most of my experience lies in this modality.

Polarity Therapy (PT) is an energy medicine system based on the idea that humans possess a biofield and the energy in their biofield can be manipulated for improved quality of life and increased health and wellness. Using touch, verbal interaction, exercise, nutrition and other methods[5], polarity practitioners seek to balance and restore the natural flow of energy in the biofield. Blocked and stagnant energy is responsible for both emotional and physical pain as well as disease. Energy medicine treatments are patient-practitioner oriented, where both the giver and receiver of the energy treatment must work in tandem for the treatment to occur. The practitioner grounds and centers his/her body, meaning all thoughts, emotions and physical sensations are neutralized through intention. This mindset begins the healing process for both the practitioner and the client. During optimal healing states, our body resonates at certain frequencies (0.3-30 Hz). PT balances the subtle energy of our biofield which can be detected and manipulated by movements of the practitioner's hands. The practitioner provides the resonating template for the client's biofield to follow. Change occurs on a spiritual or unconscious level, and most people do not feel much other than becoming very relaxed. People will often fall asleep. This mind-body state is optimal for healing and cell regeneration. After-effects of the treatment last from hours to days with feelings of calm, focus, peace and serenity. PT bases its philosophy in the traditional system of Ayurvedic medicine, which defines patterns of health as energy moving through the Five Elements of Life – Ether, Air, Fire, Water and Earth. The practice of PT focuses on the balance of these elements as the foundation of good health. It integrates philosophies of Ayurvedic medicine, hermetic or ancient Egyptian medicine, Traditional Chinese Medicine (TCM), chakra balancing and the balance of yin and yang[6]. PT understands that energy flows through the body along five pathways, enabled by positive and negative poles of the body. Five energy centers along the body represent the five elements of Ayurvedic tradition relating to different organs and functions in the body. Practitioners aim to correct disturbances and enable optimal physical, emotional as well as spiritual healing. Along with energy balancing sessions, cleansing diet and energy exercises are part of the therapy. PT has been shown to reduce cancer related fatigue[7, 8], and improve the quality of life for caregivers of dementia and Alzheimer's patients as well as improved stress reduction[9]. The majority of my patients report deep relaxation after treatment, and many experience emotional release.

Another biofield therapy is **chakra balancing**. The chakras are seven energy centers along the center of our body starting from the base of our spine to the top of our head (see figure 1).

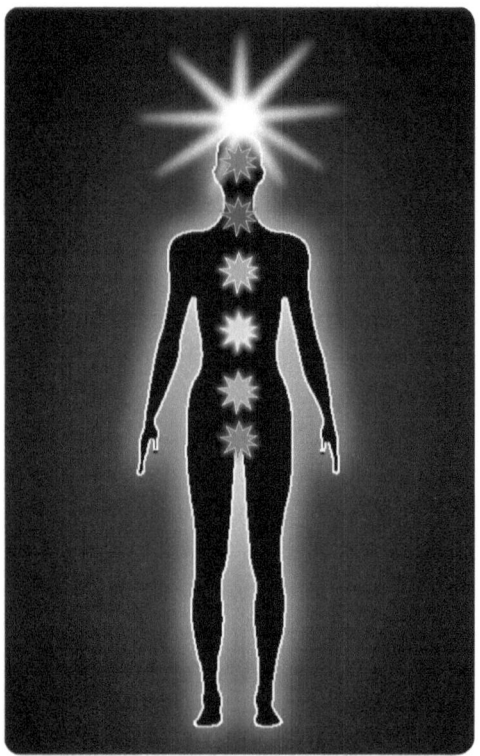

Figure 1. The location of the 7 chakras.

Chakras are our energy centers. They are responsible for keeping vital energy flowing through our biofield. They create the openings for life energy to flow into and out of our aura. Their function is to vitalize the physical body and to bring about the development of our self-consciousness. They are associated with our physical, mental, and emotional interactions. The idea that our bodies have chakras dates back thousands of years to Vedic Sanskrit writings and is an integral part of the ancient Indian healing systems of Ayurveda and yoga[10]. An energy worker trained in reading chakras will be able to determine which chakras are functioning poorly and which chakras are working too hard to keep our energy balanced. When one or two chakras are performing at a reduced level, the remaining chakras have to work harder. Having a non-functioning or closed chakra can cause another chakra to blow, creating havoc in our biofield. Blown chakras cause pain and initiate disease. Chakra balancing reprograms our chakra system to flow as nature intended.

Healing Touch (HT) is an energy medicine practice involving the relationship between the practitioner, the patient's intention, and the power of touch to facilitate healing. Like PT the patient is fully clothed, except for the removal of shoes, socks, and jewelry, is lying on a massage-type table, where the practitioner applies different forms of touch in order to evaluate the patient's energy by searching for imbalances. HT is widely used to help ease pain, stress and anxiety[11]. Patients receiving music, imagery and touch therapy during angiograms or other cardiac procedures were 65 percent less likely to die in the following six months than patients who received no such intervention[12]. Other research shows that HT lowers blood pressure, heart and breathing rates, fatigue, mood disturbances, and pain in patients receiving chemotherapy[13]. HT is endorsed by the American Holistic Nurses' Association.

Therapeutic Touch (TT) is similar to PT and HT, except practitioners usually do not actually touch the patient, but hold their hands 4-8 inches (10-18 cm) from the body in order to detect energy imbalances and correct them. TT has been shown to significantly reduce pain and increase the quality of life in fibromyalgia patients[14], shows significant improvements in pain and function in patients with osteoarthritis[15], produces significant reductions in behavioral symptoms of dementia[16], and chemical dependency in pregnant women who suffer from anxiety[17]. TT is mainly used by nurses.

A **Pulsed Electromagnetic Field (PEMF)** device was approved in 1982 by the US Food and Drug Administration (FDA) for bone repair, although it remains widely unused due to physician misunderstanding and lack of knowledge concerning the treatment[18]. PEMF therapeutic devices can be applied in two different ways - either by capacitive or inductive coupling. In capacitive coupling there is no contact with the body, whereas direct coupling requires the placement of opposing electrodes in direct contact with the surface of the skin of the targeted tissue[19]. Inductive coupling does not require electrodes to be in direct contact with the skin because it produces a field (see Faraday's Law of Induction) that emanates in all directions.

Research shows that therapeutic applications of PEMF at extra low frequency (ELF) levels (3-300 Hz) are beneficial to the immune system by suppressing inflammatory responses at the cell membrane level[20]. PEMF can pass through the skin and into the body's conductive tissue[21-23], resulting in reduced pain and the onset of edema shortly after trauma. Where edema is already present, treatment exhibits significant anti-inflammatory effects[24]. In a study of the effect of PEMF therapy on arthritis, three hours of exposure to a 50-Hz magnetic field revealed that experimentally-induced inflammation in rats was significantly inhibited as a result[25]. Strong beneficial effects have also occurred using 75 Hz frequency MF treatment in patients suffering from fractures of the ankle joints[26].

Studies show that PEMF treatments can promote cell activation and endothelial cell proliferation through the cell membrane. ELF levels can increase the rate of formation of epithelial cells in partially healed wounds[27] and also quicken the healing time of skin wounds[28]. Fields at 15 Hz were used to significantly accelerate wound healing in diabetic mice[29]. Skin wounds have electrical potentials that can be stimulated by ELF-EMF to aid in the healing process by dedifferentiating cells near the wound, thereby accelerating cell proliferation[20]. In a study examining the effects of whole body magnetic fields (50-165 Hz) on patients suffering from different forms of cancer, results showed the MF therapy had overall beneficial effects, particularly with respect to improved immune status and postoperative recovery[30]. Treatment consisted of 15 cycles, each 1-20 minutes in duration and was coupled with more traditional cancer therapies.

EMF treatments also appear to improve certain psychological conditions. A study of twelve patients with posttraumatic stress disorder (PTSD) and major depression underwent PEMF treatment of either 1 Hz or 5 Hz as an adjunct to antidepressant medications. Seventy-five percent of the patients had a clinically significant antidepressant response after treatment, and 50 percent had sustained that response at 2-month follow-up as compared with controls. Comparable improvements were seen in anxiety, hostility, and insomnia[31]. Low-frequency PEMF therapy at 0.1 – 64 Hz has been shown to improve mobility function, pain, and fatigue in fibromyalgia patients[32].

Reiki is an ancient Tibetan Buddhist practice in which practitioners serve as facilitators for life force energy (chi). Practitioners use 12 – 15 specific hand positions each held for a few minutes on the patient's clothed body. Sessions last 30-90 minutes and the number of treatments vary depending on patient response. Reiki is used to reduce stress, improve health and quality of life, and promote mental clarity. Like other energy medicine treatments, Reiki practitioners believe treatments can be effective over long distances. Formal scientific evidence has shown that Reiki can increase quality of life and reduce pain when used with standard medications[33]. Reiki has also been shown to relieve stress and improve psychological well being in patients with heart rate variability[34], and pain management[35].

In **Traditional Chinese Medicine (TCM)** the basis for disease results from the disruption in the flow of subtle energy known as qi or chi. TCM works with imbalances in the forces of yin (feminine principle) and yang (masculine principle). Practices such as Chinese herbs, meditation, massage and acupuncture aid healing by restoring yin-yang and chi or qi to homeostasis. This same subtle energy is known as ki in the Japanese Kamp system. **Acupuncture** uses meridians of Eastern medicine traditions that form a continuous, semiconducting network. It originated in China as a family of procedures that uses the stimulation of specific points on the body where the insertion of needles through the skin removes blockages in the flow of chi through the meridians in the body in order to reinstate health. During the first office visit, the practitioner asks many questions about health conditions, lifestyle, and behavior. The practitioner will want to obtain a complete picture of your treatment needs and behaviors contributing to your condition. Inform the acupuncturist about all treatments or medications you are taking and all medical conditions you have.

Acupuncture needles are metallic, solid, and hair-thin. People experience acupuncture differently, but most feel little or minimal pain while the needles are inserted. Some people feel energized by treatment, while others feel relaxed. Acupuncture has been shown to improve treatment related pain in cancer patients[36], pain management for women in labor[37], treatment of temporomandibular (TMJ) disorders[38], treat infertility, improve symptoms of menopause[39], improve insomnia[40], and improve chances of successful in vitro fertilization[41]. Relatively few complications have been reported from the use of acupuncture; however, acupuncture can cause potentially serious side effects if not delivered properly by a qualified practitioner. Make sure your practitioner is a certified Licensed Acupuncturist (L.Ac.).

Transform Your Life through Energy Medicine (TYLEM) addresses health challenges through spiritual growth and healing; promotes stress reduction and relaxation, pain management, healing relationships, coping with grief or loss, reshaping family dynamics, trauma and disaster recovery. Dr. Mary Jo Bulbrook is a spiritual/medical intuitive who practices and teaches this form of energy medicine (see http://www.energymedicinepartnerships. com)"

Craniosacral therapy is a manipulation technique that involves light touch to the cranium and sacrum (skull and base of spine and tailbone). It is based on the theory that the movement of bones within the skull and the lower back, as well as the rhythmic flow of cerebrospinal fluid (in and around the spinal cord), play a central role in the body's overall function. Obstruction of this flow of spinal fluid contributes to problems in the brain, spine and endocrine system. Research shows statistically significant improvements in the treatment of migraine headache[42].

Bowen Technique is a gentle technique involving a series of moves held for several seconds and then released. The therapist gently pulls the skin on the back of the neck, knees, or affected body part away from the muscle or tendon beneath it and applies light pressure following a specific pattern. Bowen relieves both physical and psychosocial problems, including pain, sports injuries[43], shoulder problems[44], postpartum symptoms[45], fatigue, anger and depression[46].

Trager Approach is a combination of massage, meditation and movement education. The head, torso, arms and legs are manipulated with rhythmic pull and rotation techniques in order to release tension, increase mobility and clear the mind. Movement awareness is emphasized to promote relaxation and ease neuromuscular pain. The Trager Approach has been shown to reduce symptoms of chronic headache along with reduction of headache medication[47]. I have been treated with the Trager Approach for over 20 years and find it beneficial in reducing tension and joint pain as well as promoting emotional balance and a feeling of well being.

Nambudripad's Allergy Elimination Techniques (NAET®), are a non-invasive, drug free solution to alleviate allergies using a combination of energy balancing, testing and treatment procedures from acupuncture/acupressure, allopathy, chiropractic, nutritional, and kinesiological disciplines of medicine.

Research has shown in to be effective in reducing symptoms of allergies to milk, sugar, egg white, pork meat, and other foods, causing eczema and dyspnea (shortness of breath)[48].

Brennan Healing Science is a system of energy healing that combines hands-on healing techniques with spiritual and psychological processes which include the chakras, meridians and auric field. (See www.barbarabrennan. com).

Distance and intuitive healing teaches that our biofield is full of data and information about our life, our spirit, our higher purpose, even your future. Caroline Myss is the leader in this field. (See www.myss.com/).

Rolfing focuses on fascia tissue which connects all internal structures within the human frame. Connective tissue unites the structure of our body and divides it into individual functioning parts. Fascia is constantly changing and adapting in response to demands placed on our body. It reacts to trauma – to a joint for example – by producing extra tissue to enhance stability and support. However, it can produce more than is necessary. In time, rather than stabilizing movement it can actually reduce mobility, leading to changed posture and altered patterns of movement. After completing ten sessions a client can expect greater ease of movement and all over range of motion, along with better posture. Rolfing has been shown to significantly decrease pain and increase range of motion in adults who have cervical spine dysfunction[49].

Feldenkrais Method is a form of somatic education that uses gentle movement and directed attention to improve range of motion and enhance human functioning. It can improve flexibility and coordination. The Feldenkrais Method is based on principles of physics, biomechanics and an empirical understanding of learning and human development. Feldenkrais exercises have been reported to be an effective way to improve balance and mobility, helping to offset age-related declines in mobility and reduce the risk of falling among community-dwelling older adults[50].

Alexander Technique teaches the ability to improve physical postural habits, particularly those that have become ingrained or are conditioned responses. The technique has been purported to improve performance, self observation and impulse control and relieve chronic stiffness, tension and stress. It changes movement habits in everyday activities, improving ease and freedom of movement, balance, support and coordination, teaching the use of appropriate amount of effort for a particular activity, and increasing energy. It is not a series of treatments or exercises, but rather a reeducation of the mind and body. It can be applied to sitting, lying down, standing, walking, lifting, and other daily activities. Strong evidence exists for the effectiveness of Alexander Technique lessons for chronic back pain and moderate evidence in Parkinson's-associated disability[51].

Like pharmaceuticals, the effectiveness of these treatments varies with each patient. It is important to talk with a practitioner before scheduling an appointment to discuss your needs and ask questions about what to expect during your visit. Most of these modalities are practiced with the patient fully clothed except for shoes, socks and jewelry. Wearing comfortable cotton clothes, such as yoga clothes is optimal. Although bioenergy fields have as yet not proved measurable with conventional scientific equipment, medical journals have published articles suggesting the existence of such fields[52-54].

Other forms of CAM

Aromatherapy uses aromatic plant extracts called essential oils to promote health and well-being. Works primarily through the olfactory system (sense of smell), which is our most primitive sense linked to the brain's emotional center. Used to aid relaxation, improve mood, reduce stress and anxiety, and alleviate symptoms of disease[55], it has been used successfully in the treatment of migraine headache[42].

Heart Coherence is based on heart rate variability (HRV) which is a physiological mechanism that enables our body to adapt to changing stressors. Greater variability means more flexible response to sympathetic nervous system (SNS) "fight or flight" response mechanisms. Reduced HRV has been linked to aging, trauma, and various diseases such as coronary heart disease[56, 57].

Homeopathy uses the dilutions of substances from plants, minerals and animals to treat illness. It is based on the belief that "like cures like" and the substances that cause the symptoms of illness, when highly diluted, can treat the illness. It is used by many dentists[58] and veterinarians[59] for the reduction of pain and swelling.

Meditation involves the self-regulation of attention, awareness, insight and understanding of inner and outer experiences. Practitioners claim regular meditation leads to enhanced concentration, alertness, mental efficiency and productivity. It can also lower blood pressure, heart and breathing rates, increase parasympathetic nervous system function, reduce levels of stress hormones such as epinephrine and cortisol while increasing melatonin and improve the immune system and the body's ability to heal itself[60].

Naturopathy educates the patient to prevent illness and if they get sick to support the body's capacity to heal itself. Naturopaths use nutritional supplements, botanical medicine, spinal manipulation, hydrotherapy, exercises, Traditional Chinese Medicine and Ayurveda to address disease. Naturopathy is used mainly to treat chronic and degenerative diseases such as asthma and osteoarthritis. Naturopaths treat food as medicine and stress the importance of a diet high in whole grains, fresh organic fruits and vegetables and low consumption of processed foods. It has been used extensively in detoxification[61].

Probiotics stimulate the immune system to keep potentially harmful microorganisms in check. Many of the bacteria that naturally live in the human gut aid digestion and protect the intestinal lining. These "friendly bacteria" or probiotics inhibit the growth of harmful organisms and boost our immune system[62] improving symptoms of inflammatory bowel diseases[63].

Supplements, vitamins and minerals include essential fatty acids present in fish oils, and essential amino acids, antioxidants and fiber. These protect our body against diseases such as diabetes, high blood pressure, cancer and heart disease[64,65]. Vitamins are substances derived from foods that are required for health but cannot be manufactured in the body, except for vitamin D which is produced when the skin is exposed to sunlight. Vitamins perform functions that are vital to digestion, metabolism, growth, resistance to infection and healthy muscle, nerve and tissue function. Minerals occur naturally in food and are major components of bone and teeth, the manufacturing of enzymes, and cell structure. Minerals play important roles is cellular respiration by carrying oxygen to cells and removing carbon dioxide waste. Minerals include calcium, magnesium, phosphorous, sodium, chloride, potassium, copper, iron, zinc, iodine, fluoride, chromium, selenium, manganese and molybdenum.

Thermography also known as digital infrared thermal imaging (DITI) is used as a method of research for early clinical diagnosis and control during treatment of homeostatic imbalances. It is non-invasive and uses no radiation, but instead uses the heat from our body to diagnose disease and health conditions. Using an infrared radiation camera and a computer, it measures temperatures ranging from 10° C - 55° C to detect areas that are potentials for tumor growth, neurological and physiological disease, as well as sports injury. It is commonly used for early breast cancer detection for women undergoing biopsy[66]. The medical applications of DITI are extensive, particularly in the fields of Rheumatology, Neurology, Oncology, Physiotherapy and sports medicine.

Where to start

Since the number of diseases is astronomical and there are at least 6,000 rare diseases and over 12,000 disease categories listed on the World Health Organization website, finding a CAM that is best for each individual disease can be difficult. Locating the system or body part originally affected by the disease is the best place to start. For example, diabetes would be centered on the pancreas, viruses would center on the immune system, blood diseases affect the circulatory system, etc. Next determine what element is affected by the disease. Blood is earth (red), nervous system (air), viruses and autoimmune diseases (water), brain (ether) and diseases having to do with any organ in the digestive tract (fire). Reference Table 1 to find the associated element for the disease manifested. Ears are ether, nose is earth, eyes are fire, chin and mouth are water, and skin is air. Change in diet initiates the healing process as the elements return to equilibrium (see chapter on *Food and Health*). Energy medicine stabilizes the reorganization of the field, while changes in thought processes along with spiritual focus create a healing environment.

Chakra Number	Disease	Location	Related Element	Related Color	Related Endocrine Gland
1st	Blood	Circulatory System	Earth	Red	Adrenals
2nd	Immune	Lymphatic System	Water	Orange	Sexual Organs
3rd	Liver Gallbladder Intestines Stomach	Digestive System	Fire	Yellow	Pancreas
4th	Heart	Cardiovascular System	Air	Green	Thymus
5th	Throat Esophagus	Throat	Ether	Blue	Thyroid
6th	Brain	Cranium	Ether	Violet	Pituitary Gland
7th	Memory Mental Illness	Cognitive	Ether	Magenta	Pineal Gland

Table 1. Elemental relationship of disease and body system

Whichever CAM you may be considering first consult with your physician to determine any health issues (recent surgeries or trauma) which may not allow for physical manipulation. Many energy medicine treatments are not recommended for people who have multi-personality disorder or schizophrenia as manipulation of the biofield can exacerbate delusion, hallucination and bring out all personalities at once. Be sure to mention all prescription drugs you take to determine any contraindications from combining pharmaceuticals with supplements.

References

1. Dossey, L., *Space, Time and Medicine*. Boston, MA: Shambhala Publications, Inc, 1982.
2. Woolf, S.e.a., *Promoting Informed Choice: Transforming Health Care to Dispense Knowledge for Decision Making*. Annals of Internal Medicine, 2005. **143**(4): p. 293-300.
3. Eysenbach, G., *Consumer Health Informatics*. BMJ, 2000. **320**(7251): p. 1713-6.
4. Lorenz, L., Chilingerian, J, *Using visual and narrative methods to achieve fair process in clinical care*. J Vis Exp., 2011. **16**(48): p. pii: 2342.
5. Association, A.P.T., *APTA*. Standards for Practice, 2003. **Fourth Edition**: p. 2.
6. Stone, R., *Polarity Therapy*. Summertown, TN., 1986. **2**.
7. Mustian, K., Roscoe, J, Palesh, O, Sprod, L, Heckler, C, Peppone, L, Usuki, K, Ling, M, Brasacchio, R, Morrow, G, *Polarity Therapy for Cancer-Related Fatigue in Patients With Breast Cancer Receiving Radiation Therapy: A Randomized Controlled Pilot Study*. Integr Cancer Ther., 2011. **10**(1): p. 27-37.
8. Pierce, B., *The use of biofield therapies in cancer care*. Clin J Oncol Nurs, 2007. **11**(2): p. 253-258.
9. Korn, L., Logsdon, R, Polissar, N, Gomez-Beloz, A, waters, T, Tyser, R, *A Randomized Trial of a CAM Therapy for Stress Reduction in American Indian and Alaskan Native Family Caregivers*. The Gerontologist, 2009. **32**: p. 1-10.
10. Society, S., *Chakras and kundalini yoga*. http://www.sanatansociety.org/chakras/chakras.htm, 2011.
11. Wardell, D., Weymouth, K *Review of studies of Healing Touch*. Journal of Nursing Scholarship: Image, 2004. **36**(2): p. 147-154.
12. Krucoff, M., *Healing touch, music, relaxation a plus for heart surgery patients*. The Lancet, 2005.
13. Post-White, J., Kinney, M, Savik, K, Gau, J, Wilcox, C, Lerner, I, *Therapeutic massage and healing touch improve symptoms in cancer*. Integr Cancer Ther., 2003. **2**(4): p. 332-344.
14. Denison, B., *Touch the pain away: new research on therapeutic touch and persons with fibromyalgia syndrome*. Holist Nurs Pract, 2004. **18**(3): p. 142-151.
15. Gordon, A., Merenstein, J, DiAmico, F, Hudgens, D, *The effects of therapeutic touch*

on patients with osteoarthritis of the knee. Journal of Family Practice, 1998. **47**: p. 271-277.

16. Woodsa, D., Beckb, C, Sinha, K, *The Effect of Therapeutic Touch on Behavioral Symptoms and Cortisol in Persons with Dementia.* 2009. **16**(3): p. 181-189.

17. Larden, C., Palmer, L, Janssen, P, *Therapeutic Touch Eases Anxiety for Pregnant, Chemically Dependent Women.* Journal of Holistic Nursing, 2004. **22**(4): p. 320-332.

18. Bassett, A., D. Carpenter, and S. Ayrapetyan, *Therapeutic uses of electric and magnetic fields in orthopedics.* Biological Effects of Electric and Magnetic Fields: Beneficial and Harmful Effects, 1994. **II**(San Diego: Academic Press): p. 13-48.

19. Trock, D., *Electromagnetic Fields and Magnets: Investigational Treatment for Musculoskeletal Disorders.* Rheum Dis Clin North Am., 2000. **26**(1): p. 51-62.

20. O'Connor, M., R. Bentall, and J. Monahan, *Emerging Electromagnetic Medicine conference proceedings.* Springer-Verlag, New York, 1990.

21. Stiller, M., et al., *A portable pulsed electromagnetic field (PEMF) device to enhance healing of recalcitrant venous ulcers: a double-blind placebo-controlled clinical trial.* Br J Dermatol., 1992. **27**: p. 147-154.

22. Hannan, C., et al., *Chemotherapy of Human Carcinoma Xenografts during Pulsed Magnetic Field Exposure.* Anticancer Research, 1994. **14**: p. 1521-1524.

23. Traina, G., et al., *Use of Electric and Magnetic Stimulation in Orthopaedics and Traumatology: Consensus Conference.* J Ortho Trauma, 1998. **24**(1): p. 1-31.

24. Morris, C. and T. Skalak, *Acute exposure to a moderate strength static magnetic field reduces edema formation in rats.* Am J Physiol Heart Circ Physiol, 2007. **294**: p. H50-H57.

25. Mizushima, Y., I. Akaoka, and Y. Nishida, *Effects of Magnetic Field on Inflammation.* Experientia, 1975. **31**(12): p. 1141-1412.

26. Gromak, G. and G. Lacis, *Evaluations of the efficacy of using a constant magnetic field in treatments of patients with traumas.* Electromagnetic Therapies of Injuries and Diseases of the Support-Motor Apparatus., 1987(International Collection of Papers, Riga, Latvia: Riga Medical Institute): p. 88-95.

27. Mertz, P., S. Davis, and W. Eaglestein, *Pulsed electrical stimulation increases the rate of epithlialization in partial thickess wounds.* Transactions of the 8th Annual Meeting ot the Bioelectrical Repair and Growth Society, 1988(Washington, D.C.): p. October 9-12.

28. Ottani, V., et al., *Effects of pulsed extremely-low frequency magnetic fields on skin wounds in the rat.* Bioelectromagnetics, 1988. **9**: p. 53-62.

29. Callaghan, M., et al., *Pulsed electromagnetic fields acclerate normal and diabetic wound healing by increasing endogenous FGF-2 release.* Plast Reconstr Surg. , 2007. **121**(1): p. 130-141.

30. Lubennikov, F., A. Lazarev, and V. Golubtsov, *First Experience in Using a Whole-Body Magnetic Field Exposure in Treating Cancer Patients.* Vopr Onkol, 1995. **41**(2): p. 140-141.

31. Rosenberg, P., et al., *Repetitive Transcranial Magnetic Stimulation Treatment of Comorbid Posttraumatic Stress Disorder and Major Depression.* The Journal of Neuropsychiatry and Clinical Neurosciences 2002. **14**: p. 270-276.

32. Sutbeyaz, S., et al., *Low-frequency pulsed electromagnetic field therapy in fibromyalgia:*

a randomized, double-blind, sham-controlled clinical study. Clin J Pain., 2009. **25**(8): p. 722-728.

33. Olson, K., Hanson, J, Michaud, M, *A Phase II Trial of Reiki for the Management of Pain in Advanced Cancer Patients.* Journal of Pain and Symptom Management 2003. **26**.

34. Baldwin, A., Wagers, C, Schwartz, G, *Reiki improves heart rate homeostasis in laboratory rats.* J Altern Complement Med., 2008. **14**(4): p. 417-422.

35. Lucas, M., Olson, K *Reiki To Manage Pain.* Cancer Prevention and Control, 1997. **1**: p. 108-113.

36. Lin, J., Chen, Y, *The role of acupuncture in cancer supportive care.* Am J Chin Med., 40012. **2**(219-229).

37. Jones, L., Othman, M, Dowswell, T, Alfirevic, Z, Gates, S, Newburn, M, Jordan, S, Lavender, T, Neilson, J, *Pain management for women in labour: an overview of systematic reviews.* Cochrane Database Syst Rev, 2012. **Mar 14;3:CD009234**.

38. Itoh, K., Asai, S, Ohyabu, H, Ima,i K, Kitakoji, H, *Effects of trigger point acupuncture treatment on temporomandibular disorders: a preliminary randomized clinical trial.* J Acupunct Meridian Stud, 2012. **5**(2): p. 57-62.

39. Baumelou, A., Liu, B, Wang, X, Nie, G, *Perspectives in clinical research of acupuncture on menopausal symptoms.* Chin J Integr Med., 2011. **17**(12): p. 893-7.

40. Ganguly, G., *Acupuncture may be helpful only for patients with comorbid insomnia secondary to chronic pain syndromes.* J Clin Sleep Med, 2011. **7**(4): p. 411.

41. Huang, D., Huang, G, Lu, F, Stefan, D, Andreas, N, Robert, G, *Acupuncture for infertility: is it an effective therapy?* Chin J Integr Med., 2011. **17**(5): p. 386-95.

42. Curtis, P., Gaylord, S, Park, J, Faurot, K, Coble, R, Suchindran, C, Coeytaux, R, Wilkinson, L, Mann, J, *Credibility of low-strength static magnet therapy as an attention control intervention for a randomized controlled study of CranioSacral therapy for migraine headaches.* J Altern Complement Med., 2011. **17**(8): p. 711-21.

43. James, L., *Bowen Technique for back pain and other conditions.* Positive Health 2008. **143**(38–39).

44. Carter, B., *A pilot study to evaluate the effectiveness of Bowen technique in the management of clients with frozen shoulder.* Complement Ther Med, 2001. **9**: p. 208–15.

45. Stiles, K., *Bowtech.* Massage Ther J, 2003. **42**: p. 94–104.

46. Shapiro, G., *The Bowen Technique for pain relief.* Positive Health Phys, 2004: p. 48–51.

47. Foster, K., Liskin, J, Cen, S, Abbott, A, Armisen, V, Globe, D, Knox, L, Mitchell, M, Shtir, C, Azen, S, *The Trager approach in the treatment of chronic headache: a pilot study.* Altern Ther Health Med., 2004. **10**(5): p. 40-46.

48. Terwee, C., *Succesful treatment of food allergy with Nambudripad's Allergy Elimination Techniques (NAET) in a 3-year old: A case report.* Cases J, 2008. **1**(1): p. 166.

49. James, H., Castaneda, L, Miller, M, Findley, T, *Rolfing structural integration treatment of cervical spine dysfunction.* J Bodyw Mov Ther, 2009. **13**(3): p. 229-38.

50. Ullmann, G., Williams, H, Hussey, J, Durstine, J, McClenaghan, B, *Effects of Feldenkrais exercises on balance, mobility, balance confidence, and gait performance in community-dwelling adults age 65 and older.* J Altern Complement Med., 2010. **16**(1): p. 97-105.

51. Woodman, J., Moore, N, *Evidence for the effectiveness of Alexander Technique lessons in medical and health-related conditions: a systematic review.* Int J Clin Pract., 2012. **66**(1): p. 98-112.

52. Anderson, J., Taylor, A, *Effects of healing touch in clinical practice: a systematic review of randomized clinical trials.* J Holist Nurs., 2011. **29**(3): p. 221-228.

53. Hart, L., Freel, M, Haylock, P, Lutgendorf, S, *The use of healing touch in integrative oncology.* Clin J Oncol Nurs, 2011. **15**(5): p. 519-525.

54. Mills, P., Jain, S, *Biofield therapies and psychoneuroimmunology.* Brain Behav Immun, 2010. **24**(8): p. 1229-30.

55. Lee, Y., Wu, Y, Tsang, H, Leung, A, Cheung W, *A systematic review on the anxiolytic effects of aromatherapy in people with anxiety symptoms.* J Altern Complement Med., 2011. **17**(2): p. 101-108.

56. Oakley, I., Emond, L, *Diabetic cardiac autonomic neuropathy and anesthetic management: review of the literature.* AANA J, 2011. **79**(6): p. 473-9.

57. Pragodpol, P., Ryan, C, *Critical Review of Factors Predicting Health-Related Quality of Life in Newly Diagnosed Coronary Artery Disease Patients.* J Cardiovasc Nurs, 2012. **Apr 9. [Epub ahead of print]**.

58. Horobin, D., *Useful remedies.* Br Dent J, 2011. **210**(12): p. 557.

59. Lees, P., *Pharmacology and/or homeopathy.* Vet Rec., 2011. **169**(14).

60. Merkes, M., *Mindfulness-based stress reduction for people with chronic diseases.* Aust J Prim Health, 2010. **16**(3): p. 200-10.

61. Allen, J., Montalto, M, Lovejoy, J, Weber, W, *Detoxification in naturopathic medicine: a survey.* J Altern Complement Med., 2011. **17**(12): p. 1175-80.

62. Yan, F., Polk, D, *Probiotics and immune health.* Curr Opin Gastroenterol., 2011. **27**(6): p. 496-501.

63. Meijer, B., Dieleman, L, *Probiotics in the treatment of human inflammatory bowel diseases.* J Clin Gastroenterol., 2011. **45**(Suppl:S): p. 139-44.

64. Saldeen, T., Mehta, J, *Dietary modulations in the prevention of coronary artery disease: a special emphasis on vitamins and fish oil.* Curr Opin Cardiol, 2002. **17**(5): p. 559-67.

65. Pauwels, E., *The protective effect of the Mediterranean diet: focus on cancer and cardiovascular risk.* Med Princ Pract., 2011. **20**(2): p. 103-11.

66. Wishart, G., Campisi, M, Boswell, M, Chapman, D, Shackleton, V, Iddles, S, Hallett, A, Britton, P, *The accuracy of digital infrared imaging for breast cancer detection in women undergoing breast biopsy.* Eur J Surg Oncol, 2010. **36**(6): p. 535-40.

CHAPTER NINETEEN
Living a Poor Quality of Life vs. Dying

"Though every man will attempt in his own way to postpone questions and issues related to death, until he is forced to face them, he will only be able to change things if he can start to conceive of his own death. This must be achieved by every human being alone. If each of us would make a start by contemplating the possibility of our own personal death, we may affect many things, most important of all the welfare of our patients, our families, and finally perhaps our nations."

~ Dr. Elizabeth Kubler-Ross

Disease and poor Quality of Life

Quality of Life (QOL) can only be measured and determine by each of us as individuals because each of us sees quality on a different level. Many quadriplegics feel a high QOL, while able bodied people often complain of a poor quality of life. QOL is as much a state of mind as it is measureable by a test. In cases where someone is facing a chronic or terminal illness their QOL can either deteriorate rapidly or decrease over time. If we are chronically or terminally ill, or physically or mentally disabled, then we must consider whether our quality of life has benefitted from the type of treatment we are receiving. If we are unable to determine this alone, we must rely on our family and medical practitioners. In situations where critical decisions regarding our health have to be made by health care workers, they may not be aware of how to evaluate our QOL. Some physicians and health care providers guess at QOL issues based on age, or whether the patient is bedridden, in a nursing home, or if the patient is physically handicapped or has dementia. There are formal QOL measurements that are used such as tests of physical functioning or psychological factors, but there is concern in the ethical and medical community about the use and accuracy of QOL conclusions by health care workers of their patients[1]. The question arises as to whether the QOL of a patient should be a decision-making factor in the care of the patient and if so should physicians and other health care workers make such an evaluation. Many times people are kept on life support only for the benefit of family members and loved ones who do not want to let go. It is important to discuss death and dying with our loved ones before this situation is at hand. It is also important to determine who makes the decision about when the pain and anguish of treatment is no longer worth the struggle to stay alive. Good physicians will suggest the discontinuation of chemotherapy when it is no longer affective due to its many harmful side effects. An oncologist who is honest about the effects of a treatment is a good doctor. Because oncologists are in the business of curing cancer, many times they feel a sense of failure when a patient is getting sicker, so they keep treating even though the treatment is no longer effective. There may come a time when there is nothing left to experience in life except unremitting physical, mental and emotional pain combined with a total loss of dignity. Only the patient can decide when it is time to move through the process of death. If that is not possible (patient is in a coma, or so drugged as to be incapacitated), then family and physicians must make that decision. It is time for society to move past the idea that death is failure. Life in our physical body is a gift of unspeakable proportions[2], but so is death. Death is as powerful a transition as birth. We came from someplace and we will return to that same place. Life is something we have chosen to experience and death is moving toward another way of life.

When Does a Poor Quality of Life (QOL) Become Worse Than Dying?

Death occurs because our soul is residing in a body that no longer functions due to trauma, old age, or incurable disease. Incurable disease manifests because our soul is finished with its life's work. "Our soul is clear that its purpose is evolution. It is not concerned with the achievements of the body or the development of the mind - these are all meaningless to the soul. The soul understands there is no great tragedy involved in leaving our body. In many ways the tragedy is being in the body"[2].

It is difficult to imagine that a life cut short at a young age would be desirable to any soul. Many of our young die from incurable diseases or tragic accidents. Sometimes their life is spared only to live in a vegetative state. We discuss the horrors of such conditions and pray the patient fully recovers; however we rarely find meaning in this situation. The meaning is left to be interpreted by those in the patient's life. Incurable diseases leave us exhausted, taking vital energy from every cell in our body. Treatments such as chemotherapy, high-dose pharmaceuticals or radiation often deplete the remaining energy in our body, leaving us with a poor quality of life. Is it better to live a poor quality of life or transition to our next life?

What Happens When We Die?

Death is a taboo subject in the US. Some cultures are more open to talking about death especially those who believe in reincarnation. People who have had near death experiences (NDEs) have suggested that life after death is very peaceful, serene, warm and loving[3]. They express feelings of disassociation from their body, levitation and visions of light. In his book *Home with God: in a Life that Never Ends*, Neale Donald Walsch describes death as a moment of creation. There is an energy adjustment at the moment of death that fine-tunes our biofield producing a duplicating effect in the nonphysical world of the afterlife. We continue to have the same life experience as we transition into another realm. The same process occurs at birth, only in reverse. When we are born, the energy we have brought with us from the spiritual realm is transformed into matter by the same duplicating effect of the physical world we just entered. Death is a doorway through which we experience our thought patterns. For instance if we do not believe in God we will not experience God after death. God will still be there – we just will not experience God being present. Walsch uses the analogy that if we look at a flower and know that God is in the flower then we will see God in the flower. Otherwise we will see nothing more than a flower, or perhaps a weed. If we look into the eyes of another and know that God is there we will see God there, otherwise we will see nothing more than another human being, or perhaps a villain[2].

Immediately following our passing from the physical world, every illusion of physical life is revealed to be just that – an illusion. All thoughts are energy and that includes the afterlife. Death is a process to reestablish our true identity. Heaven is not an actual place, but a state of being. It is the expression of Divinity Itself[2]. We all die when our life on earth is complete. Our life on earth is complete when we have experienced all we came here to experience. No one dies having failed to experience all they came to experience in the physical world. There is no such thing as an incomplete life. The death of every person always serves the agenda of every other person who is aware of it. *That is why they are aware of it.* Therefore, no death (and no life) is ever wasted. No one ever dies in vain[2]. All deaths bring a message to those who leave earth and to those who remain. It is for those of us who remain to find the message in the life that just left and understand its meaning. Walsch describes death as the realization we are not our body; and without it we are still very much alive. All feelings of fear or apprehension or uneasiness drop away and a flow of sweet warmth pours gently over our soul covering it entirely. Nothing about our essence is hidden, overlooked or missed. Through a non-judgmental energy that melts even the smallest sense of shame or pride, leaving our soul with a beautiful emptiness, holding nothing within, through which our soul is filled with comfort, profound appreciation, long awaited protection and unconditional love. Walsch describes this individual selfhood as a submersion into the "breathtaking glory of unending magnificence, unparalleled beauty, and unequaled completeness of being".

There are as many interpretations of life after death as there are religions and cultures to interpret it. If you are experiencing incurable disease and have not given much thought to death, perhaps now is the time to contemplate it. Thinking about death does not mean giving up on life. So many of my clients with incurable disease tell me their loved ones refuse to talk about death with them because it scares them. One of the first questions I ask my beloved patients is, "How do you feel about dying?" At first they look at me surprised, and then are relieved that someone is actually willing to talk about death with them.

For ages authors and poets have written about death and dying. James Blanchard Cisneros in his book *You Have Chosen to Remember: a Journey from Perception to Knowledge, Peace of Mind & Joy,* writes:

> "Your birth is not your beginning and neither will your death be your end. As birth is a continuation of your soul's journey, so too is death a continuation. Depending on the circumstances surrounding your passing, you might choose to remain on the physical plane for awhile. The more sudden the death and the less consciously prepared you are for it, the longer you will probably stick around. Once out of your body, you will experience a pulling sensation away from your physical body. You will feel more relaxed than afraid. The pulling sensation will give you the feeling of passing through a tunnel. So now, you are out of your body. Your body is dead,

but not your soul. You again feel that pulling sensation. You are still in control, and you can stop being pulled any time you wish. Some people choose to stay near their bodies until the funeral. However, souls really have little or no interest in their bodies once they are released. It feels as if they are looking at an old empty suitcase, they simply have no attachment to it. The main reason a soul chooses to hang around is to assist those left behind through the transition. A soul understands that it will see all those left behind soon enough. The soul will try to reach out to those in pain in order to comfort them. But this is where a problem occurs: those left behind are usually in grief, due in part to the false belief that they will never see their loved one again. Because of this grief, it is much harder for the soul to get through to these people. Sometimes, because of how these people are feeling, the only time the soul can comfort them is during sleep. In dreams, the soul will come to the individual and tell him or her that the soul is fine. If an individual is able to be calm, meditate or sit quietly, the soul will have a greater chance of consciously letting him or her know that it is fine and at peace"[4].

These are peaceful interpretations of death. Another writer's view on the subject of death and dying is from Pat Rodegast's book *Emmanuel*.

On Dying Young

"An individual dies young because he has completed his task. The being that has passed will be reminded of his task and that it was indeed his time. He will be offered a glance of the future of all those left behind, and he will find comfort and peace in this. You are an eternal being. Your brother is an eternal being. Once you leave this earth, this so-called young soul becomes a very old soul[5]."

On Preparing for the Loss of Loved Ones

"There are two answers. Loved ones are never lost. You must experience it in your own way. Of course, you will miss the physical being but when you learn to go beyond that, there will be no missing at all. Even as you sit in your human form, once you allow yourself, notice the word 'allow' to believe that you exist beyond the physical, you will touch hands with those who have left. And it will be real. It will be more real than the physicality that you had touched before. Are you aware that the physical body is a shield or a shell? It does not reveal but rather hinders revelation. If you did not have need of illusion you would not need a physical body at all"[5].

What should you do immediately following the death of a loved one?

"That is an excellent question. First, the willingness to let that person go to the next step in their evolution is extremely helpful, not only to you but to them. A 'farewell', a 'bon voyage', a 'Godspeed', then the rest of you look at each other and give comfort and assurance and provide all the hugs and tissue that are necessary. Next take yourself to a place of great luxury and enjoy an incredible feast. Salute the soul that has completed its task. Give a toast to the time you will meet again and go about the business of your own lives. Death is not only a time of mourning it is a time of truth. Karmic ties can be formed by an unwillingness to express any negativity thus holding resentments that go into the soul consciousness to return in another life. By your dealing with the negative emotion, by cleansing the relationship, you are helping both of you. The saying 'Don't speak ill of the dead', is nonsense. There is no such thing, even when you are dead, you are still alive. You do not cease to exist at death – that is only an illusion. You go through the doorway of death alive and there is no altering of your consciousness. It is not a strange land you go to but a land of living reality where the growth process is a continuation. Dying is self-regulating. It is of Divine origin. It is absolutely safe. The fear of death is the fear of letting go. As it is in life, so it is in death. The process of dying is *always* a joyous one once the human has laid aside fear. There is nothing to fear in the universe. *Nothing*. When a soul leaves its physical body, as in a profound meditation, there is a light, a sense of well-being, of peace, and of knowing that you are there in your entirety, in your individuality. You have not ceased to exist but have gone to another level of more intense existence"[4].

Emmanuel goes on to say that it is important to remain vitally alive in the decision-making process during the final act of completion of the physical life. Death is a passage through time. One of the first joys after death is the reconstruction of the image of self into the Oneness of all things without for one moment losing the Self. "Why would anyone desire to remain in physical reality when their task is finished and they are approaching Light? It is really difficult to comprehend the tenacity with which someone will cling to a decaying and useless body when such joy and Light await at the exit door. *To be vibrantly alive at your moment of death is to grow to the fullest possible extent in a lifetime*".[5]

I wanted to share these beautiful interpretations of death and dying, especially in Western culture where there is so much fear and dread surrounding death. Many countries celebrate death instead of morning it [6,7]. Of late there is a trend toward "celebration of life" where loved ones dress in color instead of black and share amusing stories of the life of the one who has passed on. Laughing through tears is a perfect blend of emotion. Great truths have been shared at these gatherings and many people brought together who had not seen each other in years have let go of animosity at funerals. The hardest part of dying is certainly letting go of those surrounding you – watching them grieve. When we give our loved ones permission to die, it is a show of unselfish, unconditional love. Keeping our loved ones alive despite a poor QOL (because we do not want to experience life without them) can be considered selfish. When we are dying it is important to discuss this with those we love, explaining how important their thoughts and feelings are to us, and how we wish to be remembered. Discussing funeral and burial plans is not morbid, but sustaining to those we leave behind because they do not have to make so many decisions after we are gone. It is also important to discuss how we feel about death and dying. Do we believe in life after death? Do we believe in a higher power? How will we say goodbye? "See you on the other side!" seems to leave them with a smile.

Elizabeth Kubler-Ross: "I say to people who care for people who are dying, if you really love that person and want to help them, be with them when their end comes close. Sit with them - you don't even have to talk. You don't have to do anything but really be there with them". In our society it is difficult to accept death because it is unfamiliar in spite of the fact that it happens all the time. We are afraid of sadness and grief, but they are natural emotions, they are part of us that allows us to say goodbye even when we are experiencing loss and do not want to say goodbye. It is not the end of the physical body that should worry us, rather our concern must be to live while we are alive - to release our inner selves from the spiritual death that comes with living behind a facade designed to conform to external definitions of who and what we are. It's only when we truly know and understand that we have a limited time on earth - and that we have no way of knowing when our time is up - we will then begin to live each day to the fullest, as if it was the only one we had".

I have lost loved ones to cancer, diabetes and other terminal illnesses. These losses have shifted my thinking about my life and ultimately my death. After leaving their funeral I set aside time for contemplation – for focusing inward. I reevaluate my life and how I am living it with respect to those still on earth. Death brings about change. Death is a release, an end to pain and suffering. Death is a transformation…it's going home.

References

1. Hannon, B., O'Reilly, V, Bennett, K, Breen, K, Lawlor, P, Meeting the family: Measuring effectiveness of family meetings in a specialist inpatient palliative care unit. Palliat Support Care, 2012. **10**(1): p. 43-49.
2. Walsch, N., Home with God: in a life that never ends. Atria Books, 2006. **New York, NY**.
3. Cant, R., Cooper, S, Chung, C, O'Connor, M, The divided self: near death experiences of resuscitated patients--a review of literature. Int Emerg Nurs, 2012. **20**(2): p. 88-93.
4. Cisneros, J., You Have Chosen to Remember: A Journey from Perception to Knowledge, Peace of Mind and Joy. Cisneros Capital Group: Miami, FL, 2004: p. 287.
5. Rodegast, P., Stanton, J, Emmanuel. Bantam Books, New York, 1985.
6. Neimeyer, R., Meaning Reconstruction and the Experience of Loss. American Psychological Association:Washington DC 2001.
7. Barley, N., Grave Matters: A Lively History of Death around the World. New York: Henry Holt and Company, 1997.

CHAPTER TWENTY
How Can Disease Change My Life For the Better?

Disease is a gift inasmuch as we have opportunity to experience both our vulnerability and power at the same time. Illness changes not only our life, but the lives of those around us. Because of the fear we have of dying it is difficult to see a bigger picture, but our relationships will change, for better or worse, they will change. With change comes opportunity for growth. When my loved ones are diagnosed with a terminal illness, I turn to guidance during these times of deep sorrow as my emotions prevent me from seeing the good in things that at first appear horrible. I refer to literary artists and poets whose words fill the void where mine fail. The book *Emmanuel* is a must read for anyone experiencing disease or facing death. The wisdom contained in this book is rare. Here is an excerpt from Pat Rodegast's book:

> "So much can be gathered in that time of quietness, of introspection, that illness forces upon you dear souls who are always outer motivated. Such times can be used for the alchemy of taking the clay of physicality and breathing the spirit into it that will change that clay into gold. *Illness is a teaching, a message from the soul. When the lessons are learned the illness becomes a thing of no moment".*

Disease is the physical manifestation of confusion in our soul so we become consciously aware of the imbalance and frustration our soul is experiencing. Embracing illness, whether it is mental, physical or emotional is the beginning of transformation. Pain speaks to us when we are ready to learn from it. Emotional pain says one thing, physical pain another. Even its location in our body is eloquent. Nothing in life happens haphazardly. I realize this is a hard thing to hear when someone is in pain, but the truth is the truth. We live in a sane and ordered universe. We must make that our tenet[1].

> *Illness exists first in the non-physical realm of spiritual need, emotional confusion, or mental aberration. It is never primarily physical. The body is the reactor. It vibrates to stress and is an outward manifestation of inner turmoil.* As the body constricts under the onslaught of trauma, there is denial of energy to a particular part of the body. Thus the stage is set for a physical manifestation which is a malfunction of the body.
>
> *-Emmanuel*

In conventional medicine illnesses are classified by their symptoms, but their causes can be totally different. The exact same illness may exist in two different people for two different reasons. Illness is the body's expression of disunity. Spontaneous healing can take place if there is recognition of the truth behind what has caused the disease; therefore illness is the unwillingness of the conscious mind to perceive the expression of our soul. As our life force or soul consciousness flows through our physical body, those areas that resist the life force energy can develop a dysfunction at some point depending on the needs of our soul[1].

> *Your body in illness is not your enemy but your faithful friend. It has been programmed by your soul to react in that exact way at that exact time. Do heed its guidance".* - Emmanuel

When asked if illnesses such as cancer and HIV/AIDS are a plague from God, Emmanuel replies:

> "Oh, my dears, what a terrible thought that God would ever send a plague. How can we hope to heal this affliction if we step aside and blame it on a deity? No God of loving compassion would ever inflict upon anyone, be it individual, group or community, any illness at all, not even the common cold. If you believe that you are deserving of punishment you need to question what you feel is amiss in your life. Is there

a sense of guilt? Where you align yourself with any posture that hints you are deserving of God's punishment there lies the issue at hand, not the illness itself".

When asked if mentally ill people are consciously in control of their sanity, Emmanuel responds that they are not mentally in control of their sanity, but their spirit is in control of their sanity. To say that one controls one's insanity is to make a cruel statement. Many times mental illness is the commitment of our soul, for not just the sake of our own growth, but for the contribution to the learning experience of others.

With respect to cancer – it is an issue of fear. Cancer brings the message of fear that is so prevalent in our world, so it must be dealt with squarely as fear. Once cancer is cured there will be another disease until we begin to deal with our fear, because fear is the denial of the reality of God. Fear is the opposite of love - not hate. Hate is based in fear. Until we overcome our fear there will be no cure for cancer[1].

With respect to acquired immune deficiency syndrome (AIDS) – it is an issue of judgment. AIDS bring with it the message of condemnation that is so prevalent in our society. It questions acceptance of others. Acceptance of all people is essential to understanding that life is divine. We are all part of the same God. Until we accept all cultures, races, religions and lifestyles, there will be no cure for AIDS.

With respect to heart disease – it is an issue of love. Heart disease brings with it the message of unconditional love. It questions our understanding of love from a spiritual perspective. To be able to love our self the same way God loves us is essential to our health and happiness. Until we truly understand unconditional love of self and of others there will be no cure for heart disease.

With respect to diabetes – it is an issue of unworthiness. Diabetes brings the message of poor self esteem that is so prevalent in our society. It questions whether we are good enough. It refuses to see God in humanity. Embracing our divinity is imperative to staying healthy, vital and alive. Until we see the worthiness of our life there will be no cure for diabetes.

Be aware that all diseases can be cured if one is wise enough to consider death a cure. So many in our society see death as failure, but death is the passing of one life form into another. When our soul is ready to leave our body we could be walking around healthy as ever and be hit by a bus. If our soul is not ready to pass over we will recover – our body will heal itself[1]. Our body is innately designed to regenerate each and every part. There is power in our body's ability to heal itself, but insisting that we "must" heal is overlooking an opportunity. Living peacefully with disease is enlightening.

References

1.   Rodegast, P., Stanton, J, Emmanuel. Bantam Books, New York, 1985.

CHAPTER TWENTY ONE
Conclusion

The manifestation of disease provides us with the information that our soul is not completing its life's work. We are given this information in our psychological and emotional body before it manifests as physical health problems. There are many available orthodox medical treatments for mental and physical disease, but the ability to pay for such treatments is often controlled by outside influences. If we are empowered to take charge of our own health by either preventing disease or increasing our quality of life despite disease, it is imperative that we ask questions, research all options and determine the best possible health care plan for our condition. What is best may include alternative or complementary options to current methods of treatment. Understanding the human organism is more than the sum of its systems, organs, cells and molecules - it is an information system known as the human biofield that brings with it many options not often recommended by conventional health care practitioners. Treatments can be received both by professionals or trained family members depending on the financial situation. Many of the healing practices can be taught to family members during one or two weekend sessions. Actively participating in one's own healing process is important not only to the patient, but the patient's family and friends. Feelings of helplessness in facing terminal illness give rise to depression, hopelessness and despair. Healing arts practiced by friends and family creates loving relationships that previously may not have existed.

It is as important to focus on the *prevention* of disease as it is to treat disease. Disease prevention would decrease soaring health care costs present in the current political strife of today's health care crisis. Our current health care model focuses not on prevention, but repair for accidents and trauma, and offers surgery and pharmaceuticals to ease symptoms of diseases whose etiology is unknown. In 2009 the US spent $2.5 trillion for an average of $8,047 per person on health care[1]. In a report studying the collaboration of eight hospitals to determine if comprehensive lifestyle changes and treatments offered by integrative and functional medicine practitioners could be a safe and effective alternative to bypass surgery or angioplasty, nearly 80 percent of participants were able to safely avoid heart surgery or angioplasty and the calculated savings reached almost $30,000 per patient in the first year alone[2].

The political battle on health care reform offers few if any changes to a system that needs a complete overhaul. Only a grassroots level of involvement will change this mismanaged network. The practice of integrative medicine provides a comprehensive health care model of practical, patient-based solutions to many of the chronic diseases straining our unsustainable process of health care. Chronic and incurable illnesses like heart disease, diabetes, obesity and cancer account for more than half of all health care costs and more than 70 percent of all deaths. The CDC maintains that lack of physical activity, poor nutrition, tobacco use and excessive alcohol consumption are responsible for much of the pain and suffering related to chronic illness and disease[3]. As health care costs rise and quality of life declines, choices must be made by the public, not congress, not "Big Pharma" nor the medical community, but by those experiencing a decline in health and financial status due to disease. The use of patient-centered integrative medicine in combination with individualized, scientifically based functional medicine would allow both patients and their health care providers to more accurately identify early stages of chronic disease and prevent the suffering and high cost of later stage illness. There is a better way. It is up to us as individuals to insist that we, not money, not big business, not the government, *we* decide how to proceed. May you enjoy health and happiness for the remainder of your days.

Always,

Chris

References

1. Jones, B., *Medical expenses have 'very steep rate of growth.* USA Today, 2010. http://www.usatoday.com/news/health/2010-02-04-health-care-costs_N.htm(February 4, 2010).
2. Ornish, D.e.a., *The Lifestyle Heart Trial.* The Lancet, 1990. **336**(8708): p. 129-33.
3. Prevention., N.C.f.D.C.a., *Chronic Disease Prevention and Health Promotion.* Centers for Disease Control and Prevention, 2010. http://www.cdc.gov/chronicdisease/overview/index.htm.

CHAPTER TWENTY TWO
Resources

CAM Therapy	Treatment Research	Find a Practitioner
Acupressure	Improves cognitive brain function after traumatic brain injury[1] Relieves pain[2] Reduces tension and stress[3] Prevents nausea[4] Treats obesity[5] Provides cardiovascular improvement after stroke[6] Offsets cancer related fatigue[7]	www.acupressurenearyou.com
Acupuncture	Improves treatment related pain in cancer patients[8] Pain management for women in labor[9] Treatment of TMJ disorders[10], Treats infertility and improves chances of successful in vitro fertilization[11] Improve symptoms of menopause[12] Improves insomnia[13].	www.acufinder.com/ Acupuncture+Associations

CAM Therapy	Treatment Research	Find a Practitioner
Aromatherapy	Aids relaxation, improves mood, reduces stress and anxiety, alleviates symptoms of disease[14]	www.naha.org US www.ifparoma.org UK cfacanada.com Canada
Ayurveda	Improves asthma symptoms[15] Improves angina symptoms[16]	www.acufinder
Bowen Technique	Increases range of motion in inflamed joints[17] Stroke rehabilitation[18] Low Back Pain[19]	www.thebowentechnique.com www.bowen-therapy.co
Chiropractic	Chest pain[20] Osteoarthritis[21] Cyclic vomiting[22] Migraine Headaches[23] Scoliosis[24] Neck pain[25]	http://www.acatoday.org/ US http://www.chiropractic.org/Int'l
Energy Psychology	Thought field therapy (TFT) Trauma (PTSD)[26] Whole Health Easily Effectively (WHEE) Test Anxiety[27] Emotional Freedom Technique (EFT) fibromyalgia[28] Tapas Acupressure Technique (TAT) Weight Loss[29]	www.energypsych.org/
Heart coherence (HRV)	Chronic fatigue syndrome[30] Ischemic Heart Disease[31]	http://www.heartmath.org
Herbal remedies	Too many to list here.	http://nccam.nih.gov/research

CAM Therapy	Treatment Research	Find a Practitioner
Healing touch	Pain, stress and anxiety[32] After cardiac procedures[33] Chemotherapy symptoms[34, 35]	www.healingtouchinternational.org/
Homeopathy	Atopic dermatitis[36] Attention-deficit/ hyperactivity disorder (ADHD)[37] Rheumatoid Arthritis[38]	http://nationalcenterforhomeopathy.org/
Meditation	Alzheimer's Disease[39] Anxiety Disorder[40] Self-Control Depletion[41] Phantom Limb Amputation[42]	www.learning-modern-meditation.com/ how-to-practice-mindfulness.html
Naturopathy	Sinusitis[43] Dysmenorrhea[44] Acne[45]	www.naturopathic.org/
Nambudripad's Allergy Elimination Techniques (NAET)	eczema and dyspnoe in reaction to food allergy[46]	www.naet.com
Osteopathy	Low Back Pain[47, 48]	www.academyofosteopathy.org
PEMF therapy	Non-union bone fractures[49, 50] Diabetic Wound Healing[51] Skin lymphoma[52]	www.pemft.org/
Polarity Therapy	Cancer related fatigue[53, 54] Dementia[55]	www.polaritytherapy.org/

CAM Therapy	Treatment Research	Find a Practitioner
Probiotics	Depleted immune system[56] Irritable bowel syndrome[57]	www.usprobiotics.org/
Reiki	Pain and anxiety in the elderly with dementia[58] Work related stress of nursing[59] Depression[60]	http://www.reiki.org/
Therapeutic Touch(TT)	Fibromyalgia[61] Pain and anxiety in burn patients[62] Neurological complications during bone marrow transplant[63]	www.councilforhealing.org/TT.html
Traditional Chinese Medicine (TCM)	Cancer symptoms[64] Side effects of chemo[65] Migraine headache[66] Pain and inflammation[67] Cerebral ischemia[68]	www.atcm.co.uk/ www.nccaom.org/
Trager Approach	Chronic headache[69]	www.trager.com/approach.html
Yoga	Osteoarthritis[70] Pain[71] Stress in school age girls[72] Psychiatric Disorders[73]	www.abc-of-yoga.com/ yogabeginnersguide.asp

References

1. McFadden, K., Healy, K, Dettmann, M, Kaye, J, Ito, T, Hernández, T, *Acupressure as a non-pharmacological intervention for traumatic brain injury (TBI)*. J Neurotrauma., 2011. **28**(1): p. 21-34.

2. Smith, C., Collins, C, Crowther, C, Levett, K, *Acupuncture or acupressure for pain management in labour*. Cochrane Database Syst Rev, 2011. **Jul 6;**(7): p. CD009232.

3. Chang, K., Wong, T, Wong, T, Leung, A, Chung, J, *Effect of acupressure in treating urodynamic stress incontinence: a randomized controlled trial*. Am J Chin Med. , 2011. **39**(6): p. 1139-1159.

4. Suh, E., *The effects of p6 acupressure and nurse-provided counseling on chemotherapy-*

induced nausea and vomiting in patients with breast cancer. Oncol Nurs Forum, 2012. **39**(1): p. E1-9.

5. Hsieh, C., Su, T, Fang, Y, Chou, P, *Effects of auricular acupressure on weight reduction and abdominal obesity in Asian young adults: a randomized controlled trial.* Am J Chin Med., 2011. **39**(3): p. 433-440.

6. McFadden, K., Hernández, T, *Cardiovascular benefits of acupressure (Jin Shin) following stroke.* Complement Ther Med, 2010. **18**(1): p. 42-48.

7. Zick, S., Alrawi, S, Merel, G, Burris, B, Sen, A, Litzinger, A, Harris, E, *Relaxation acupressure reduces persistent cancer-related fatigue.* Evid Based Complement Alternat Med, 2011. **pii: 142913**(Sept 2).

8. Lin, J., Chen, Y, *The role of acupuncture in cancer supportive care.* Am J Chin Med., 40012. **2**(219-229).

9. Jones, L., Othman, M, Dowswell, T, Alfirevic, Z, Gates, S, Newburn, M, Jordan, S, Lavender, T, Neilson, J, *Pain management for women in labour: an overview of systematic reviews.* Cochrane Database Syst Rev, 2012. **Mar 14;3:CD009234**.

10. Itoh, K., Asai, S, Ohyabu, H, Ima,i K, Kitakoji, H, *Effects of trigger point acupuncture treatment on temporomandibular disorders: a preliminary randomized clinical trial.* J Acupunct Meridian Stud, 2012. **5**(2): p. 57-62.

11. Huang, D., Huang, G, Lu, F, Stefan, D, Andreas, N, Robert, G, *Acupuncture for infertility: is it an effective therapy?* Chin J Integr Med., 2011. **17**(5): p. 386-95.

12. Baumelou, A., Liu, B, Wang, X, Nie, G, *Perspectives in clinical research of acupuncture on menopausal symptoms.* Chin J Integr Med., 2011. **17**(12): p. 893-7.

13. Ganguly, G., *Acupuncture may be helpful only for patients with comorbid insomnia secondary to chronic pain syndromes.* J Clin Sleep Med, 2011. **7**(4): p. 411.

14. Lee, Y., Wu, Y, Tsang, H, Leung, A, Cheung W, *A systematic review on the anxiolytic effects of aromatherapy in people with anxiety symptoms.* J Altern Complement Med., 2011. **17**(2): p. 101-108.

15. Joos, e.a., *Immunomodulatory effects of acupuncture in the treatment of allergic asthma: a randomized controlled study.* Altern Complement Med, 2000. **6**(6): p. 519-25.

16. Chao, e.a., *Nalaxone reverses inhibitory effect of electroacupuncture on sympathetic cardiovascular reflex responses.* American Journal of Physiology, 1999. **45**(6): p. H1227-H2134.

17. Carter, B., *A pilot study to evaluate the effectiveness of Bowen technique in the management of clients with frozen shoulder.* Complement Ther Med, 2001. **9**: p. 208–15.

18. Duncan, B., McHugh, P, Houghton, F, Wilson, C, *Improved motor function with Bowen therapy for rehabilitation in chronic stroke: a pilot study.* JOURNAL OF PRIMARY HEALTH CARE, 2011. **3**(1): p. 53-57.

19. Marr, M., *The effects of the Bowen technique on hamstring flexibility over time: A randomized controlled trial.* Journal of Bodywork and Movement Therapies, 2011. **15**(3): p. 281 – 290.

20. Stochkendahl, M., Christensen, H, Vach, W, Høilund-Carlsen, P, Haghfelt, T, Hartvigsen, J, *Chiropractic Treatment vs Self-Management in Patients With Acute Chest Pain: A Randomized Controlled Trial of Patients Without Acute Coronary Syndrome.* J Manipulative Physiol Ther., 2012 **35**(1): p. 7-17.

21. Strunk, R., Hanses, M, *Chiropractic care of a 70-year-old female patient with hip osteoarthritis.* J Chiropr Med, 2011. **10**(1): p. 54-59.

22. Hubbard, T., Crisp, C, *Cessation of cyclic vomiting in a 7-year-old girl after upper cervical chiropractic care: a case report.* J Chiropr Med. , 2010. **9**(4): p. 179-183.

23. Chaibi, A., Tuchin, P, *Chiropractic spinal manipulative treatment of migraine headache of 40-year duration using Gonstead method: a case study.* J Chiropr Med., 2011 **10**(3): p. 189-193.

24. Morningstar, M., *Outcomes for adult scoliosis patients receiving chiropractic rehabilitation: a 24-month retrospective analysis.* J Chiropr Med. , 2011. **10**(3): p. 179-184.

25. Dunn, A., Green, B, Formolo, L, Chicoine, D, *Chiropractic management for veterans with neck pain: a retrospective study of clinical outcomes.* J Manipulative Physiol Ther., 2011. **34**(8): p. 533-538.

26. Diepold, J., Goldstein, D, *Thought field therapy and QEEG changes in the treatment of trauma: a case study.* Traumatology, 2008. **5**: p. 1.

27. Benor, D., Ledger, K, Toussaint, L, Hett, G, Zaccaro, D, *Pilot study of emotional freedom technique (EFT), wholistic hybrid derived from EMDR and EFT (WHEE) and cognitive behavioral therapy (CBT) for treatment of test anxiety in university students.* Explore, 2009. **5**(6): p. 1.

28. Brattberg, G., *Self-administered emotional freedom techniques (EFT) in individuals with fibromyalgia: a randomized trial.* Integrative Medicine: a clinicians journal, 2008. **august/september.**

29. Elder, C., Ritenbaugh, C, Mist, S, Aickin, M, Schnieder, J, Zwickey, H, Elmer, P, *Randomized trial of two mind-body interventions for weight-loss maintenance.* j of altern and complem med, 2007. **13**(1): p. 67-78.

30. McCraty, R., Lanson, S, Atkinson, M, *Assessment of Autonomic Function and Balance in Chronic Fatigue Patients Using 24-Hour Heart Rate Variability Analysis.* Clinical Autonomic Research, 1997. **7**(5): p. 237.

31. Umetani, K., Singer,D, McCraty, R, Atkinson, M, *Cycle Length Dependency of Heart Rate Variability in Elderly with Ischemic Heart Disease.* Circulation (suppl I), 1996. **94**(8): p. I-498.

32. Wardell, D., Weymouth, K *Review of studies of Healing Touch.* Journal of Nursing Scholarship: Image, 2004. **36**(2): p. 147-154.

33. Krucoff, M., *Healing touch, music, relaxation a plus for heart surgery patients.* The Lancet, 2005.

34. Post-White, J., Kinney, M, Savik, K, Gau, J, Wilcox, C, Lerner, I, *Therapeutic massage and healing touch improve symptoms in cancer.* Integr Cancer Ther., 2003. **2**(4): p. 332-344.

35. Lutgendorf, S., Mullen-House, E, Russell, D, et al, *Preservation of immune function in cervical cancer patients during chemoradiation using a novel integrative approach.* Brain, Behavior, and Immunity, 2010. **24**: p. 1231-1240.

36. Eizayaga, J., Eizayaga, J, *Prospective observational study of 42 patients with atopic dermatitis treated with homeopathic medicines.* Homeopathy. , 2012 **101**(1): p. 21-27.

37. Pellow, J., Solomon, E, Barnard, C, *Complementary and alternative medical therapies for children with attention-deficit/hyperactivity disorder (ADHD).* Altern Med Rev., 2011. **16**(4): p. 323-37.

38. Brien, S., Leydon, G, Lewith, G, *Homeopathy enables rheumatoid arthritis patients to cope with their chronic ill health: A qualitative study of patient's perceptions of the homeopathic consultation.* Patient Educ Couns., 2012 Dec;89(3):507-16.
39. Innes, K., Selfe, T, Brown, C, Rose K, Thompson-Heisterman, A, *The effects of meditation on perceived stress and related indices of psychological status and sympathetic activation in persons with Alzheimer's disease and their caregivers: a pilot study.* Evid Based Complement Alternat Med, 2012:927509.
40. Lu, C., Smith, L, Gau, C, *Exploring the Zen meditation experiences of patients with generalized anxiety disorder: a focus-group approach.* J Nurs Res., 2012. 20(1): p. 43-51.
41. Friese, M., Messner, C, Schaffner, Y, *Mindfulness meditation counteracts self-control depletion.* Conscious Cogn, 2012 Jun;21(2):1016-22.
42. Moura, V., Faurot, K, Gaylord, S, Mann, J, Sill, M, Lynch, C, Lee, M, *Mind-Body Interventions for Treatment of Phantom Limb Pain in Persons with Amputation.* Am J Phys Med Rehabil. , 2012 Aug;91(8):701-14.
43. Kraft, K., *Naturopathy consultation. Sinusitis.* MMW Fortschr Med, 2011. 153(40): p. 19.
44. Beer, A., *Naturopathy consultation. Dysmenorrhea.* MMW Fortschr Med, 2011. 153(42): p. 23.
45. Bacharach-Buhles, M., Beer, A, *Naturopathy consultation. Acne.* MMW Fortschr Med, 2011. 153(14): p. 18.
46. Terwee, C., *Succesful treatment of food allergy with Nambudripad's Allergy Elimination Techniques (NAET) in a 3-year old: A case report.* Cases J, 2008. 1(1): p. 166.
47. Licciardone, J., Brimhall, A, King, L, *Osteopathic manipulative treatment for low back pain: a systematic review and meta-analysis of randomized controlled trials.* BMC Musculoskelet Disord, 2005. 6: p. 43.
48. Chou, R., Huffman, L *Nonpharmacologic therapies for acute and chronic low back pain: A review of the evidence for an American Pain Society/American College of Physicians clinical practice guideline.* Annals of internal medicine, 2007. 147(7): p. 492–504.
49. Holmes, G., *Treatment of delayed unions and nonunions of the proximal fifth metatarsal with pulsed electromagnetic fields.* Foot Ankle Int, 1994. 15: p. 552-556.
50. Salzman, C., Lightfoot, A, Amendola, A, *PEMF as treatment for delayed helaing of foot and ankle arthrodesis.* Foot Ankle Int, 2004. 25: p. 771-773.
51. Callaghan, M., et al., *Pulsed electromagnetic fields acclerate normal and diabetic wound healing by increasing endogenous FGF-2 release.* Plast Reconstr Surg. , 2007. 121(1): p. 130-141.
52. Rodin, I., Lamotkin, I, Ushakov, A, Skvortsov, S, Velichko, A, Nekrasova, O, Lazarenko, E, *The use of a low-intensity eddy-current magnetic field in treating patients with skin lymphomas.* Voen Med Zh., 1996. 317(12): p. 32-34.
53. Mustian, K., Roscoe, J, Palesh, O, Sprod, L, Heckler, C, Peppone, L, Usuki, K, Ling, M, Brasacchio, R, Morrow, G, *Polarity Therapy for Cancer-Related Fatigue in Patients With Breast Cancer Receiving Radiation Therapy: A Randomized Controlled Pilot Study.* Integr Cancer Ther., 2011. 10(1): p. 27-37.
54. Pierce, B., *The use of biofield therapies in cancer care.* Clin J Oncol Nurs, 2007. 11(2): p. 253-258.
55. Korn, L., Logsdon, R, Polissar, N, Gomez-Beloz, A, waters, T, Tyser, R, *A Randomized Trial of a CAM Therapy for Stress Reduction in American Indian and Alaskan Native Family Caregivers.* The Gerontologist, 2009. 32: p. 1-10.

56. Yan, F., Polk, D, *Probiotics and immune health.* Curr Opin Gastroenterol., 2011. **27**(6): p. 496-501.

57. Meijer, B., Dieleman, L, *Probiotics in the treatment of human inflammatory bowel diseases.* J Clin Gastroenterol., 2011. **45**(Suppl:S): p. 139-44.

58. Meland, B., *Effects of Reiki on pain and anxiety in the elderly diagnosed with dementia: a series of case reports.* Altern Ther Health Med, 2009. **15**(4): p. 56-57.

59. Cuneo, C., Cooper, M, Drew, C, Naoum-Heffernan, C, Sherman, T, Walz, K, Weinberg J,, *The Effect of Reiki on Work-Related Stress of the Registered Nurse.* J Holist Nurs., 2011 Mar;29(1):33-43.

60. Shore, A., *Long-term effects of energetic healing on symptoms of psychological depression and self-perceived stress.* Altern Ther Health Med, 2004. **10**(3): p. 42-48.

61. Denison, B., *Touch the pain away: new research on therapeutic touch and persons with fibromyalgia syndrome.* Holist Nurs Pract, 2004. **18**(3): p. 142–51.

62. Turner, J., Clark, A, Gauthier, D, Williams, M, *The effect of therapeutic touch on pain and anxiety in burn patients.* J Adv Nurs, 1998. **28**: p. 10-20.

63. Smith, M., Reeder, F, Daniel, L. Baramee, J, Hagman, J, *Outcomes of touch therapies during bone marrow transplant.* Altern Ther Health Med, 2003. **9**: p. 40-49.

64. Meng, Z., Kay, Garcia, M, Hu, C, Chiang, J, Chambers, M, Rosenthal, D, Peng, H, Wu, C, Zhao, Q, Zhao, G, Liu, L, Spelman, A, Lynn Palmer, J, Wei, Q, Cohen, L, *Sham-controlled, randomised, feasibility trial of acupuncture for prevention of radiation-induced xerostomia among patients with nasopharyngeal carcinoma.* Eur J Cancer, 2012. **Jan 27**.

65. Smith, M., Bauer-Wu, S, *Traditional chinese medicine for cancer-related symptoms.* Semin Oncol Nurs, 2012 **28**(1): p. 64-74.

66. Molsberger, A., *The role of acupuncture in the treatment of migraine.* CMAJ, 2012 **Jan 9**.

67. Yuan, H., Chen, R, Huang, D, Wang, Y, Wang, W, *Analgesic and anti-inflammatory effects of balance acupuncture on experimental scapulohumeral periarthritis in rabbits.* Zhongguo Zhen Jiu, 2011 **31**(12): p. 1106-1110.

68. Yang, Z., Chen P, Yu, H, Luo, WS, Wu, Y, Pi, M, Peng, J, Liu, Y, Zhang, S, Gou, Y, *Research advances in treatment of cerebral ischemic injury by acupuncture of conception and governor vessels to promote nerve regeneration.* Zhong Xi Yi Jie He Xue Bao, 2012 **10**(1): p. 19-24.

69. Foster, K., Liskin, J, Cen, S, Abbott, A, Armisen, V, Globe, D, Knox, L, Mitchell, M, Shtir, C, Azen, S, *The Trager approach in the treatment of chronic headache: a pilot study.* Altern Ther Health Med., 2004. **10**(5): p. 40-46.

70. Ebnezar, J., Nagarathna, R, Yogitha, B, Nagendra, H, *Effect of integrated yoga therapy on pain, morning stiffness and anxiety in osteoarthritis of the knee joint: A randomized control study.* Int J Yoga, 2012. **5**(1): p. 28-36.

71. Büssing, A., Ostermann, T, Lüdtke, R, Michalsen, A, *Effects of yoga interventions on pain and pain-associated disability: a meta-analysis.* J Pain., 2012. **13**(1): p. 1-9.

72. White, L., *Reducing stress in school-age girls through mindful yoga.* J Pediatr Health Care, 2012. **26**(1): p. 45-56.

73. Cabral, P., Meyer, H, Ames D, *Effectiveness of yoga therapy as a complementary treatment for major psychiatric disorders: a meta-analysis.* Prim Care Companion CNS Disord, 2011. **13**(4): p. 8.

Glossary of Terms

ALTERNATIVE MEDICINE—a variety of therapeutic or preventive health care practices that are not typically taught or practiced in traditional medical communities and offer treatments that differ from standard medical practice.

AMYGDALA—a ganglion of the limbic system adjoining the temporal lobe of the brain and involved in emotions of fear and aggression.

ANGIOGENESIS—development of blood vessels either in the embryo or in the form of neovascularization or revascularization.

ANTIBODY—any of the numerous Y-shaped protein molecules produced by B cells as a primary immune defense. Each molecule has its own unique binding site that can combine with the complementary site of a foreign antigen such as a virus or bacterium, thereby disabling the antigen and signaling other immune defenses.

ANTIGEN—a substance that when introduced into the body stimulates the production of an antibody. Antigens include toxins, bacteria, foreign blood cells, and the cells of transplanted organs.

ASTRAL PLANE—the fourth level of the human biofield. It connects the multidimensional realms of past lives and past experiences. At this level, the act of forgiveness creates a healing response.

ATOMIC ORBITAL—a mathematical function that describes the wavelike behavior of either one or a pair of electrons in an atom. It can be used to calculate the probability of finding any electron of an atom in any specific region around the atom's nucleus.

BEAM SPLITTER—an optical device that splits a beam of light in two. Often used in holography.

BIOFIELD—a holistic or global organizing field that distributes information throughout an organism and is central to its integration by regulating its biochemistry and physiology.

BIOLOGICAL TISSUE—the ensemble of similar cells from the same origin that together carry out a specific function.

BIOPHOTONS—ultraweak photon emissions of biological systems known as weak electromagnetic waves in the optical range of the electromagnetic field spectrum.

CELL COMMUNICATION—any of several ways in which living cells of an organism communicate with one another, whether by direct contact between cells or by means of chemical signals carried out by neurotransmitter substances, hormones, and cyclic adenosine monophosphates.

CELL DIFFERENTIATION—the process by which a less specialized cell becomes a more specialized cell type. Differentiation occurs numerous times during the development of a multicellular organism as the organism changes from a simple zygote to a complex system of tissues and cell types.

CELL SIGNALING—part of a complex system of communication that governs basic cellular activities and coordinates cell actions.

CHAKRA—energy vortex that connects the human soul to its mental and physical body.

CHAKRA SYSTEM—vital energy centers that run from the base of the spine to the top of the head centered on the spinal column (root, sacral, solar plexus, heart, throat, third eye, and crown). Chakras unite body, mind, and soul in the function of physical existence.

CHI—the vital energy force believed in Taoism to be inherent in all things. The unimpeded flow of chi is understood to be essential to good health in traditional Chinese medicine.

COMPLEMENTARY AND ALTERNATIVE MEDICINE (CAM)—characterized by its focus on the whole person as a unique individual, on the energy of the body and its influence on health and disease, on the healing power of nature and the mobilization of the body's own resources to heal itself, and on the treatment of the underlying causes, rather than symptoms, of disease.

DEOXYRIBONUCLEIC ACID (DNA)—a nucleic acid that carries genetic information in a cell and is self-replicating. Consisting of two long chains of nucleotides twisted into a double helix that determines individual hereditary characteristics.

DEVELOPMENTAL BIOLOGY—a field of study of all changes associated with an organism as it progresses through the life cycle.

DOSHAS—bioenergies of vata, pitta, and kapha condensed from the five elements of ether, air, fire, water, and earth, according to the principle of Ayurveda. Responsible for the physical and emotional tendencies in the mind and body that help determine an individual's physical and mental characteristics. Imbalances in the doshas are the cause of disease.

ELECTROMAGNETIC FIELD (EMF)—a property of space caused by the motion of an electric charge. A stationary charge will produce only an electric field in the surrounding space. If the charge is moving, a magnetic field is also produced. An electric field can be produced also by a changing magnetic field.

ELECTROMAGNETIC RADIATION—a form of energy emitted and absorbed by charged particles, which exhibits wavelike behavior as it travels through space.

ENERGY FIELD—a system of transmitters and receivers of vibration that program information into the body from its environment.

ENERGY MEDICINE—a branch of complementary and alternative medicine that studies the science of medical and therapeutic applications of subtle energies.

Putative energy medicine is based on the idea that human beings are able to influence subtle forms of our body's energy with their hands, intentions, or meditation. *Veritable* energy medicine uses mechanical vibration (sound) and electromagnetic radiation (light) in order to effect health and healing.

ENTANGLEMENT—a term used in quantum theory to describe the way particles of energy/matter can be predicted to interact with each other regardless of how far apart they are.

ENTROPY—the measure of the unavailable energy in a closed thermodynamic system that is also usually considered to be a measure of the system's disorder or degree of disorder or uncertainty in a system.

EPIGENETICS—the study of heritable changes in gene expression or cell phenotype caused by mechanisms other than changes in the underlying DNA sequence. Exogenous influences of genetic expression.

ETHERIC PLANE—first or lowest layer in the human energy field or biofield.

ETIOLOGY—the branch of medicine that deals with the causes or origins of disease.

FREE RADICALS—atoms, molecules, or ions with unpaired electrons or open shell configurations that play a large role in vascular tone and blood pressure.

GENES—a hereditary unit consisting of a sequence of DNA that occupies a specific location on a chromosome and determines a particular characteristic in an organism.

HEALING TOUCH—a noninvasive method of healing that was derived from an ancient laying-on-of-hands technique, where the practitioner alters the patient's energy field through an energy transfer that moves from the hands of the practitioner to the patient.

HIPPOCAMPUS—a curved elongated ridge that extends over the floor of the descending horn of each lateral ventricle of the brain, involved in forming, storing, and processing memory.

HOLOGRAPHY—a three-dimensional image reproduced from a pattern of interference produced by a split coherent beam of radiation.

HUMAN GENOME PROJECT—an international scientific research project with the primary goal of determining the sequence of chemical base pairs which make up DNA and of identifying and mapping the approximately 20,000–25,000 genes of the human genome from both a physical and functional standpoint.

IMMUNE SYSTEM—a complex network of interacting cells, cell products, and cell-forming tissues that protects the body from pathogens and other foreign substances by destroying infected and malignant cells and removing debris. This system includes the thymus, spleen, lymph nodes, lymph tissue, stem cells, white blood cells, antibodies, and lymphokines.

INFLAMMATION—response to cellular injury marked by redness, swelling, pain, tenderness, heat, and limited function of an area of the body that serves as a mechanism to eliminate noxious agents and damaged tissue.

KAPHA—one of the three doshas in Ayurveda, condensed from the elements water and earth. It is the principle of stabilizing energy; governs growth in the body and mind; is concerned with structure, stability, lubrication, and fluid balance; and is eliminated from the body through the urine.

KIRLIAN PHOTOGRAPHY—a photographic process that records electoral discharges emanating from living objects producing an aura-like glow surrounding the object onto a photographic plate or film in which the object has direct contact.

LIVING MATRIX—the collective whole of an organism, including its energies, tissues, cells, systems, and that which connects them.

MARMA POINTS—vital energy centers which store energy for use when needed.

MERIDIANS—natural energy pathways that affect every organ and physiological system in the body.

MICROBES—a microorganism, especially a bacterium that causes disease.

MORPHOGENESIS—the biological process that causes an organism to develop its shape. It is one of three fundamental aspects of developmental biology along with the control of cell growth and cellular differentiation.

MORPHOGENETIC FIELD—a group of cells able to respond to discrete, localized biochemical signals leading to the development of specific morphological structures or organs.

NADI SYSTEM—considered channels or tubes in the human body that carry prana or chi (known as energy or life force). These energy channels can be found in both the physical and the subtle body.

NATIONAL CENTER FOR COMPLEMENTARY AND ALTERNATIVE MEDICINE (NCCAM)—part of the National Institutes of Health (NIH), this center conducts and supports basic and applied research and training and disseminates information on complementary and alternative medicine to practitioners and the public.

NATIONAL INSTITUTES OF HEALTH (NIH)—an agency in the Department of Health and Human Services whose mission is to employ science in the pursuit of knowledge to improve human health. It is the principal biomedical research agency of the federal government.

NEAR-DEATH EXPERIENCES (NDEs)—experiences reported by people on the threshold of death. Characteristics include hearing oneself pronounced dead, feelings of peacefulness, the sense of leaving one's body, the sense of moving through a dark tunnel toward a bright light, a life review, the crossing of a border, and meetings with other spiritual beings, often deceased friends and relatives. Near-death experiences are reported by about one-third of those who come close to death.

NEUROTRANSMITTERS—endogenous chemicals that transmit signals from a neuron to a target cell across a synapse.

NONSTEROIDAL ANTI-INFLAMMATORY DRUGS (NSAIDs)—medications used to control pain and inflammation.

ORGANELLES—a differentiated structure within a cell, such as a mitochondrion, vacuole, or chloroplast that performs a specific function.

ORGAN—a relatively independent part of the body that carries out one or more special functions. Examples of organs include the heart, lungs, and liver.

PATHOGEN—an agent that causes disease, especially a living microorganism such as a bacterium or fungus.

PITTA—one of the three doshas in Ayurveda condensed from the elements fire and water. It is the principle of transformation energy and governs heat and metabolism in the body; is concerned with the digestive, enzymatic, and endocrine systems; and is eliminated from the body through sweat.

POLARITY THERAPY—a holistic, energy-based system that includes bodywork, diet, exercise, and lifestyle counseling for the purpose of restoring and maintaining proper energy flow throughout the body. The underlying concept of polarity therapy is that all energy within the human body is based in electromagnetic force and that disease results from improperly dissipated energy.

PRINCIPLE OF CORRESPONDENCE—this principle embodies the Kybalion truth that there is always a correspondence between the laws and phenomena of the various planes of Being and Life. The old Hermetic axiom ran: "As above, so below; as below, so above." The universe emanates from the same Source; and the same laws, principles, and characteristics apply to each phenomenon in its own plane.

PSYCHONEUROIMMUNOLOGY (PNI)—the study of the interactions between psychological factors, the central nervous system, and the immune function as modulated by the neuroendocrine system.

PULSED ELECTROMAGNETIC FIELD (PEMF) treatment—a type of electromagnetic therapy in which small electrical currents are intermittently applied to the body.

QUALITY OF LIFE (QOL)—personal satisfaction (or dissatisfaction) with the cultural or intellectual conditions under which one lives (as distinct from material comfort).

QUANTUM ELECTRODYNAMICS (QED)—quantum field theory of the interactions of charged particles with the electromagnetic field. It describes mathematically not only all interactions of light with matter, but also those of charged particles with one another.

QUANTUM MECHANICS—the branch of physics developed in the first part of the twentieth century that was highly successful in explaining the behavior of atoms, molecules, and nuclei. Combined with the general and special theory of relativity, it revolutionized the field of physics. The new concepts were the particle properties of radiation, the wave properties of matter, the quantization of physical properties, and the idea that one can no longer know exactly where a single particle such as an electron is at any one instance.

REGENERATIVE MEDICINE—the process of regenerating human cells, tissues, or organs to restore normal function. Regenerating damaged tissues and organs in the body by replacing damaged tissue and/or by stimulating the body's own repair mechanisms to heal previously irreparable tissues or organs. Regenerative medicine also enables scientists to grow tissues and organs in the laboratory and implant them when the body cannot heal itself.

REIKI—a form of therapy that uses simple hands-on, no-touch, and visualization techniques, with the goal of improving the flow of life energy in a person.

RELIGION—the taking of another's belief systems as one's own. The acceptance of a previously established belief system in lieu of one's own inner experience. Many religions have organized behaviors, clergy, and a definition of what constitutes adherence or membership, holy places, and scriptures.

RESONANCE—the prolongation and intensification of sound produced by transmission of its vibrations to a cavity, which include organic chemical structures that cannot be accurately represented by a single structural formula differing only by electron positions.

RIBOSOME—a minute round particle composed of RNA and protein found in the cytoplasm of living cells and serves as the site of assembly for polypeptides encoded by messenger RNA.

SELF-EFFICACY—a person's belief about his or her ability and capacity to accomplish a task or to deal with the challenges of life.

SOURCE—God; First Cause; Supreme Being; Universal Life Force.

SPIRIT—the principle of conscious life; the vital principle in humans, animating the body or mediating between body and mind.

SPIRITUALITY—the concept of an ultimate reality; an inner path enabling a person to discover the essence of his/her being. Spiritual practices, including meditation, prayer, and contemplation, are intended to develop an individual's inner life. Spiritual experiences can include being connected to a larger reality, joining with other individuals or the human community, with nature or the cosmos, or with the divine realm. Spirituality is often experienced as a source of inspiration or orientation in life.

SPONTANEOUS HEALING (also known as spontaneous remission)—an unexpected improvement or cure from a disease that appears to be progressing in its severity.

SUPERCONDUCTING QUANTUM INTERFERENCE DEVICE (SQUID)—a very sensitive detector of magnetic flux. Most SQUIDs are incorporated into whole-head systems for magnetoencephalography (MEG)—the detection of magnetic fields produced by the brain.

SUPERSTRING THEORY—a mathematical theory for describing the properties of fundamental particles, which represents the particles as one-dimensional string-like objects. The properties of fundamental particles in string theory and their manner of interaction with each other depend upon the modes of vibration of the strings.

TOUCH THERAPY—a noninvasive method of healing that was derived from an ancient laying-on-of-hands technique. In TT, the practitioner alters the patient's energy field through an energy transfer that moves from the hands of the practitioner to the patient.

TRANSCUTANEOUS ELECTRICAL NERVE STIMULATION (TENS)—the use of electric current produced by a device to stimulate the nerves for therapeutic purposes.

UNIVERSAL LIFE FORCE—the life force or living energy that sustains life of organs, cells, tissues, and blood. As a spiritual entity, the life force is considered our connective flow to the creator.

VATA—one of the three doshas in Ayurveda condensed from the elements air and space. It is the principle of kinetic energy in the body; is concerned with the nervous system and with circulation, movement, and pathology; and is eliminated from the body through defecation.

VIBRATIONAL FREQUENCIES—the number of waves that pass a fixed point per unit time; also, the number of cycles or vibrations undergone in unit time by a body in periodic motion (ex: The frequency with which earth rotates is once per twenty-four hours). Frequency is expressed in units called hertz (Hz). One hertz is equal to one cycle per second.

WAVE FUNCTION—a mathematical function used in quantum mechanics to describe the propagation of the wave associated with any particle or group of particles.

YANG—the masculine active principle in nature that in Chinese cosmology is exhibited in light, heat, or dryness and that combines with yin to produce all that comes to exist.

YIN—the feminine passive principle in nature that in Chinese cosmology is exhibited in darkness, cold, or wetness and that combines with yang to produce all that comes to exist.

Index

A

acquired immune deficiency syndrome (AIDS), 1, 6, 139, 184–85, 198, 218–19. *See also* human immunodeficiency virus (HIV)
afterlife, 211, 241
alcoholism, 11, 139, 241
alpha-carotene, 166, 241
alternative medicine, 21, 58, 173–74, 187, 231, 233, 236, 241–42
alternative therapies, 41, 174–75, 241
Alzheimer's disease, 9, 31, 43, 45, 87, 109, 118, 139, 148, 164, 168, 241
American Holistic Nurses Association, 197, 241
American Hospital Association (AHA), 173, 241
American Medical Association (AMA), 185, 241
amygdala, 140–41, 148, 231, 241
angiogenesis, 76
animal tissues, 106
antibiotics, 7, 47, 60, 176, 241
antidepressants, 20, 198, 241, 243
antigens, 7, 231, 241
antioxidants, 165, 168, 203, 241
anxiety, 15, 29–30, 33, 42–45, 47, 82–83, 139–41, 143, 148–49, 178, 181, 183, 185, 197, 199

apigenin, 168, 241
apoptosis, 7, 119, 129, 241
appetite, 139–40, 148, 241
aromatherapy, 37, 202, 241–42
Artificial Happiness: The Dark Side of the New Happy Class (Dworkin), 21
Ashtanga Yoga, 45, 241
astral plane, 74, 231, 241
atomic orbital, 103, 231, 241

B

Babbitt, Edwin D., 63
balance, 15, 42, 44–45
Bancroft, Anne, 35
Baule, Gerhard, 64
beam splitter, 132, 232, 241
Becker, Robert O., 64
    The Body Electric, 109
Berger, Hans, 64
beta-cryptoxanthin, 166, 241
Bikram Yoga, 45, 241
biofield, 13, 29–31, 44, 46–49, 65, 67, 77, 82–83, 87, 89–90, 101, 129–30, 132–34, 195, 197
    levels, 74–75
    See also energy field
biological tissue, 106
Biology of Belief, The (Lipton), 12, 241, 246
biophotons, 46, 77, 155, 232, 241
birth, 3, 89, 92, 210–12

Blueprint for Immortality: The Electric Patterns of Life (Burr), 63

body
astral, 94, 99–101, 108, 112
emotional, 15, 99, 112, 140, 157, 221
etheric, 94, 99–102, 110
mental, 81, 94, 112
physical, 2, 5, 13, 47, 73, 75, 77–79, 89–90, 94–96, 98–105, 108, 110–12, 134, 140, 214
spiritual, 2–3, 87, 93–94, 100, 112
Body Electric, The (Becker), 109, 241–42
body mass index (BMI), 185
breast cancer, 9, 29–30, 167, 242
breathing, 23, 36–38, 45–46, 48, 82, 142–43, 150, 174, 178, 181, 183–85, 197, 202
bromelain, 167, 242
Burr, Harold Saxton, 63–64
Blueprint for Immortality: The Electric Patterns of Life, 63–64, 241

C

cancer, 1, 6–7, 9–10, 15, 25, 32, 38, 42, 117–18, 127, 154–55, 177, 179–80, 193, 218–19
Cancer Treatment Centers of America (CTCA), 187, 242
capsaicin, 167, 242
Cardiovascular Disease (CVD), 140, 163, 177–78, 242
caregivers, 5, 37, 44–46, 142, 196, 242
carotenoids, 166–67, 242
catechins, 168, 242
causal body. *See* body, spiritual
cell
communication, 10, 100, 104, 115–18, 121–22, 133, 179
differentiation, 10, 105, 125, 127, 129
growth, 100–101, 105, 117, 125, 163, 166, 168, 179, 236
signaling, 10, 66, 104–5, 107, 115, 117–21, 126–27, 135, 149, 162
See also cell communication

cells, 8–10, 12–13, 30, 32, 48, 63–64, 66, 76–77, 83–84, 89–91, 105–6, 149, 163, 179–80, 184
progenitor, 128
stem, 10, 89–90, 126, 128–29, 131, 135, 235
Centers for Disease Control and Prevention (CDC), 177, 222
chakra balancing, 196–97
chakras, 46, 75, 78, 95–100, 112, 142, 153, 157, 163–64, 196–97, 201, 232, 242
chakra system, 78, 197, 232, 242
chemotherapy, 44, 161, 178–79, 187, 193, 197, 210–11, 242
chi (life energy), 42, 48, 93, 96, 100–101, 108, 112, 199
chlorophyll, 167, 242
cholesterol, 118, 162–63, 166–67, 177–78, 185–86, 242
chronic obstructive pulmonary disease (COPD), 6, 87, 168, 177, 182–83
chronic respiratory disease, 177, 182–83, 242
Cisneros, James Blanchard
You Have Chosen to Remember: A Journey from Perception to Knowledge, Peace of Mind & Joy, 212
complementary and alternative medicine (CAM), 20, 41, 173–76, 193, 195, 202, 204–5
acupressure, 41–42, 242
acupuncture, 41–42, 45, 47, 62, 102, 121, 174–75, 199–200, 242
Alexander Technique, 201–2, 242
aromatherapy, 37, 202
Ayurveda, 41–42, 81, 87, 178, 197, 203, 233, 235, 237, 240, 242
Bowen Technique, 62, 121, 200, 242
Brennan Healing Science, 62, 201, 242
chiropractic, 41, 43, 121, 174, 200, 242
craniosacral therapy, 200
distance and intuitive healing, 201
Feldenkrais Method, 201, 242

healing touch (HT), 43, 62, 66, 121, 178, 197

heart coherence, 202
   See also heart rate variability (HRV)

homeopathy, 41, 43, 62, 174–75, 202, 243

massage therapy, 43–44, 174, 243

meditation, 6, 13, 23, 25, 31, 33, 36, 142, 150, 156, 174, 178, 180, 199–200, 202

Nambudripad's Allergy Elimination Techniques (NAET), 62, 200

naturopathy, 174, 202–3

osteopathy, 41, 44, 243

polarity therapy (PT), 44, 62, 66, 82, 94, 99, 121, 178, 195

probiotics, 203, 243

pulsed electromagnetic field (PEMF) therapy, 44–45, 58, 62, 119–21, 197–98

Reiki, 62, 66, 121, 178, 199, 238, 243

Rolfing, 201, 243

supplements, vitamins, and minerals, 203

thermography, 203
   See also digital infrared thermal imaging (DITI)

traditional Chinese medicine (TCM), 41, 45, 62, 91, 178, 196, 199, 203

Trager Approach, 62, 121, 200, 243

Yoga, 3, 6, 23–24, 31, 33, 42, 45–46, 150, 156, 174–75, 178, 180, 197, 202, 243

Yoga meditation, 45–46, 243

connective tissue, 45, 64, 90, 95, 107, 201

constipation, 21, 30, 139, 163–64, 166, 243

coronary artery disease (CAD), 178

cortisol, 20, 22, 110, 147–48, 202, 243

creativity, 98, 156, 180, 243

Crohn's disease, 82, 166

CRP (C-reactive protein), 118, 243

cystic fibrosis, 9, 116, 243

cytokines, 119, 135, 147–49, 162–63, 243

D

death, 12, 15, 37, 47–48, 87, 89, 116–17, 140–43, 157, 177, 181–82, 186, 209–17, 219, 236

dementia, 31, 44, 46, 66, 87, 163–64, 168, 196–97, 210, 243

deoxyribonucleic acid (DNA), 6–10, 12, 76, 90, 95, 100, 102, 111, 119–20, 127, 133–34, 166, 233–35, 243

depression (*see also* antidepressants), 2–3, 11–12, 20–21, 31, 33, 42–43, 46–47, 74, 140–43, 148–49, 178–79, 181, 183, 185, 187

developmental biology, 125–27, 233, 236, 243, 245, 248

diabetes, 6, 22, 43, 82–83, 116, 141, 177, 180–81, 184, 186, 203–4, 216, 219, 222

diet (*see also* eating habits), 6, 31, 48, 81–82, 164, 169, 175, 178, 183, 186, 194, 196, 203–4, 237, 243

digital infrared thermal imaging (DITI), 76, 203

disease, 1–3, 5–7, 9–10, 13, 29–37, 47–49, 63–64, 74–76, 89, 109–11, 115–18, 148–49, 177–78, 203–4, 217–22
   chronic, 1, 47, 84–87, 176–77, 187, 222
   genetic, 8–9, 116, 120
   language of, 6, 33–35
   phases of prevention, 19
   physiological, 1, 6, 163, 176, 203
   prevention, 19, 23–24, 115, 173, 221
   See also illness

"DNA phantom," 76, 243

doshas, 81–82, 87, 99, 233, 235, 237, 240, 243. *See also* kapha; pitta; vata

Dossey, Larry, 193

drug dependence, 47, 243

drugs, 2, 21, 24, 33–34, 47–48, 140, 154, 161–62, 184, 200, 205, 243, 247–48

Duke, Patty, 35

Dworkin, Ronald

Artificial Happiness: The Dark Side of the New Happy Class, 21, 241, 243

E

earth, 29, 43, 74, 81–84, 89, 96, 98, 163–66, 168, 181, 195, 204, 211–13, 216, 243–44
eating habits, 11, 24, 47, 135, 143, 148, 161, 163, 243–44
edema, 29, 44, 58, 85, 167, 198
Eden Energy Medicine, 62, 244
ego (self), 15, 33, 74, 86, 93, 167, 244
Einthoven, Willem, 64, 244
electrocardiogram (ECG/EKG), 14, 58, 64, 244
electroencephalogram (EEG), 14, 58, 64, 244
ELECTROMAGNETIC FIELD (EMF), 58, 65, 76–77, 107, 120, 130–31
electromagnetic radiation (EMR), 58, 64, 77, 125, 233–34, 244
elements in health and healing, 81, 163–64, 195–96
    air, 87
    earth, 83
    ether, 87
    fire, 85–87
    water, 84, 87
ellagic acid, 168, 244
Emmanuel (Rodegast), 213, 215, 217–19
emotions, 1–3, 5–6, 15, 19–21, 30–31, 37–38, 41–42, 46–47, 74–75, 77–78, 99–101, 104, 134–35, 147–49, 215–17
endocrine glands, 96, 98, 204, 244
energetic homeostasis, 30
energy field, 13, 15, 43, 48, 57, 63, 65–67, 74–79, 121, 130, 132, 135, 181, 233, 241
energy level, 67, 103, 169, 244
Energy medicine (EM), 33, 41, 46, 48–49, 57–58, 65–67, 103, 120–21, 132, 142, 181–83, 186, 195, 204
energy medicine (EM), putative, 62, 65

energy psychology (EP)
Energy psychology (EP), 46–47, 62, 142–43, 187, 244
    Emotional Freedom Technique (EFT), 62, 142–43
    tapas acupressure technique (TAT), 46, 62
    thought field therapy (TFT), 46–47, 244
entanglement, 90, 234, 244
entropy, 93, 132, 234, 244
epigenetics, 10–12, 154, 234, 244
epithelial tissues, 107
etheric field, 74, 100, 102, 244. *See also* energy field
etiology, 2, 19, 148, 154, 222, 234
exercise, 9, 11, 13, 21, 24
extracellular matrix, 64, 107

F

fear, 2, 15, 23, 31, 35, 82–83, 140–41, 149, 166, 178, 183, 193, 212, 214–15, 219
fibromyalgia, 31, 42–43, 59, 62, 142–43, 197, 199, 244
Field, The, 90, 95
fight or flight (response mechanism), 22–23, 104, 110, 149, 202
flavonoids, 166, 168, 244
food, 24
    air, 164–65, 167–68
    earth, 164–66, 168
    ether, 165, 168
    fire, 164–65, 167
    water, 164–67
    yang, 24
    yin, 24
Food and Drug Administration (FDA), 44, 62, 197
free radicals, 163, 166
Fröhlich, Herbert, 63

G

ganglia, 96, 107, 231, 244
genes, 8–12, 116, 118–20, 127, 130–31, 154, 234–35, 244

Genetic defects, 8–9, 244
Genetic Testing, 8, 244
Gerber, Richard, 90, 93, 100
glutamate, 117, 244
God (*see also* Source; Universal Life
    Force), 29, 91, 98, 156–57, 168, 178,
    211, 218–19, 239, 244–45, 248

H

happiness, 36–37, 48, 75, 149, 156, 219
Hathayogapadikpa, 95, 245
health
    emotional, 139–40, 142–43, 163, 187
    physical, 23, 67, 139, 142, 149, 156,
        221
    spiritual, 150, 153, 156
Health Forum, 173, 245
Health maintenance organizations
    (HMOs), 173, 245
heart disease, 1, 21–22, 25, 38, 82, 116,
    118, 140, 165, 168, 177–78, 187,
    202–3, 219, 222
heart rate variability (HRV), 62, 199, 202,
    242, 245
"Higher Self," 93, 245. *See also* body,
    spiritual
highly active antiretroviral therapy
    (HAART), 184, 245
hippocampus, 20, 141, 234, 245
Hippocrates (father of medicine), 65, 245
histone, 130, 245
holography, 102, 125–26, 132–34, 232,
    235, 245
homeostasis, 30, 67, 106, 109, 119, 148,
    161, 199, 245
Home with God: In a Life that Never Ends
    (Walsch), 211
Human Genome Project, 9, 235, 245
human immunodeficiency virus (HIV),
    184
Hunt, Valerie, 31
    Infinite Mind, 79
Huntington's disease, 110, 116
hyperlipidemia, 186, 245

hypertension, 32, 42, 82, 177, 182, 186–87,
    245
hypothalamic-pituitary-adrenal (HPA)
    axis, 147–48, 245

I

illness, 2, 15, 25, 32, 37, 42–43, 45, 49, 63,
    76, 102, 104, 106, 202, 217–19
    acute, 2, 6, 15, 21, 30–31, 33, 47, 176
    chronic, 2, 15, 21, 25, 30, 47, 139, 141,
        222
    emotional, 143
    mental, 153, 219
    physical, 20, 31, 49, 100, 142–43
immune system, 5–7, 13, 23, 30, 38, 106,
    111, 118–20, 140, 147, 149, 154–55,
    166, 184, 202–4
    macrophages, 7, 118–20, 184, 245
    monocytes, 118, 184, 245
    T-cells, 7, 118, 148, 184, 245
Indian Journal of Clinical Biochemistry, 43,
    245
indoles, 168, 245
Infinite Mind (Hunt), 79
inflammation, 63, 66, 86, 118–20, 148–49,
    162, 166–67, 182, 198
insomnia, 20, 30, 42, 82, 139, 141, 174, 199,
    245
Institute of Medicine (IOM), 194
insulin, 116–17, 162, 178, 180–81, 245
insurance, 173–76
International Journal of Geriatric Psychiatry,
    46, 245
International Society for the Study
    of Subtle Energy and Energy
    Medicine, 58
irritable bowel syndrome (IBS), 6, 33, 42,
    166, 245
ischemic cardiomyopathy, 178, 245

K

Kaiser Permanente, 173, 245
kapha, 43, 81–82, 99, 233, 235, 243, 245
Keller, Helen, 35
Kirlian, Semyon Davidovich, 76

Kirlian photography, 76, 235, 245
Kshurika-Upanishad, 95
Kubler-Ross, Elizabeth, 215

L

Levin, Michael, 131
"The Wisdom of the Body: Future
Techniques and Approaches
to Morphogenetic Fields
in Regenerative Medicine,
Developmental Biology, and
Cancer," 127
lifestyle, 24, 45, 81–82, 110, 139, 142, 163,
169, 175, 177, 186, 194, 199, 219,
222
limonoids, 167
linoleic acid, 162, 246
lipid oxidation, 165, 246
Lipton, Bruce
The Biology of Belief, 12, 241, 246
Living Matrix, 95, 235, 246
love, 5, 48, 74–75, 78, 98, 134, 153–54,
157, 178, 180–81, 185–86, 212, 215,
219, 246
low-density lipoprotein (LDL), 163, 166,
246
lutein, 167, 246
luteolin, 168, 246
lycopene, 165–66, 246

M

macrobiotics, 24, 246
magnetic encephalograph (MEG), 77
magnetic resonance imaging (MRI), 58,
103, 141
marmas, 99, 246
McDougall, Duncan, 157, 246
McFee, Richard, 64
meditation, 3, 6, 13, 23, 25, 31, 33, 36
melanocortin system, 148
membrane potential, 106–7
meridians, 46, 75, 78, 95–96, 101–2, 107–
9, 112, 199, 201, 235, 246
meridian system, 100–102, 108
microbes, 6–7, 107, 119, 236, 246

bacteria, 6–7, 106, 119, 154–55, 203,
231, 246
fungi, 6–7, 155, 237, 246
protozoa, 6–7, 246
viruses, 6–7, 106, 119, 179, 204, 246
mind, 1–3, 5–6, 23–24, 35, 37–38, 41–43,
45–46, 65, 74–75, 93–94, 133–34,
139–40, 147–50, 153, 155–56
minerals, 43, 162, 202–3, 243, 246
Miracle Worker, The (film), 35
mitempfindung, 101, 246
mitochondria, 105
molecular biology, 120, 127, 147, 246
molecules, 7–8, 13, 32, 48, 57, 61, 63–64, 75,
77, 91, 103–5, 117, 127, 148–49, 166
Molecules of Emotion (Pert), 148, 246–47
morphogenesis (morphogenetic biofield
instruction), 90, 125–27, 130–31,
133, 236, 246
morphogenetic field, 90, 126, 128, 133, 135
Morphogens, 127, 246
"mountaintop experience," 29
MSCs (mesenchymal stem cells), 125,
128–29, 246
multiple sclerosis, 60, 109–10, 116–17, 246
muscle tissue, 106–7
Myss, Caroline, 201

N

nadis, 95–96, 99, 112, 246
nadi system, 95
National Cancer Institute, 66, 246
National Health Interview Survey
(NHIS), 174
National Hospice and Palliative Care
Organization, 66
National Institutes of Health (NIH),
57–58
nature, 15, 48, 74–76, 81, 161, 197
near-death experiences (NDEs), 211
Nei Ching, 92, 246
nerve cells, 107, 109, 116. *See also* ganglia
neural tissue, 107
neurotransmitters, 117, 236, 246
Nobel Prize, 63–64, 246

nonsteroidal anti-inflammatory drugs
(NSAIDs), 47
North American Nursing Diagnosis
Association, 66

O

obesity, 11, 30, 42–43, 47, 82, 162, 167, 177,
180, 185–87, 222, 246
obsession, 48, 246
omega-3, 162
organelles, 95, 104–5, 116, 237, 246
organs, 10, 15, 32, 42, 44–45, 77–78, 93,
95–96, 101–2, 105–9, 111, 115–16,
126–32, 204, 235–38
organ systems
cardiovascular system, 108, 110
circulatory system, 108, 117, 130, 204
digestive system, 82, 108, 139
endocrine system, 95, 98, 108, 139, 200
integumentary system, 57, 108
lymphatic system, 108
muscular system, 108
nervous system, 22–23, 33, 36, 43,
46, 57, 64, 79, 95–96, 98, 101,
107–12, 130–32, 147, 149
autonomic (ANS), 22, 109–11
central (CNS), 22–23, 98, 100,
107, 109–10, 147, 166, 237
parasympathetic (PNS), 22–23,
33, 104, 110–11, 149, 202
sympathetic (SNS), 22–23, 33,
104, 110–11, 147–49, 202
reproductive system, 108
respiratory system, 108
skeletal system, 108
Oschman, James, 95, 247
osteoarthritis, 6, 30, 43, 60–62, 84, 166,
197, 203, 247
oval fields, 98–99
oxytocin, 150, 247

P

pain, 217
emotional, 29, 148, 210
physical, 139–40, 148, 195

painkillers, 1, 31, 247
pancreas, 98, 108, 116, 180, 204
passion, 20, 36–37, 48, 181
pathogens, 7, 107, 118–20, 148, 235, 237,
247
PEMF Knee Device, 61, 247
PEMF Mats, 62, 247
Pert, Candace, 12, 104
Molecules of Emotion, 148
phagocytes, 119, 247
pharmaceuticals (*see also* drugs), 1, 7, 22,
29, 41–42, 47, 57, 120, 142, 161, 176,
186–87, 194, 202, 205
pitta, 43, 81–82, 99, 233, 237, 243, 247
pluripotent stem cells, 128
Polarity Principle, The (Sills), 94
polyphenol, 166, 247
Popp, Fritz-Albert, 77, 155, 247
post-traumatic stress disorder (PTSD), 47,
143, 179, 198
prana. See chi (life energy)
prayer, 33, 62, 75, 154–56, 174, 180, 239, 247
Principle of Correspondence, 102, 247
Prozac, 21, 247
psychoneuroimmunology (PNI), 23, 147,
237, 247
purpose, 1–2, 5–6, 15, 20, 36–37, 48,
74–75, 89, 143, 153, 156–57, 181,
201, 210

Q

qi. See chi (life energy)
Qi gong, 62, 121
Quality of Life (QOL), 1–2, 25, 33, 43–44,
47, 84–87, 141, 143, 150, 155, 157,
169, 199, 209–11, 221–22
quantum electrodynamics (QED), 77, 238,
247
quantum mechanics, 90, 238, 240, 247
Quantum physics, 75, 77, 90, 247

R

radiation, 42–44, 58, 64, 76, 125, 161,
178–79, 187, 193–94, 203, 211, 235,
238, 247

regeneration, 45, 60, 64, 66, 104, 106, 127, 130–31, 134–35, 149, 195, 247
resilience, 143, 247
resonance, 75, 79, 91, 103, 110, 121, 131–32, 134, 161
"resonance specificity," 103
"resonant frequency," 103
resveratrol, 166, 247
rheumatoid arthritis, 43, 60, 82, 141, 247
ribonucleic acid (RNA), 7–8, 90, 127, 134, 238, 248
ribosomes, 105, 238, 248
Rodegast, Pat
    Emmanuel, 213, 217, 244, 248
Russian Academy of Sciences, 76, 248

S

schizophrenia, 10–12, 42, 148, 150, 205, 248
Schrödinger, Erwin
    What is Life? The Physical Aspect of the Living Cell, 132, 248
Self-efficacy, 183
self-esteem, 20, 29, 143, 168, 186, 219, 248
serenity, 104, 111, 149, 157, 195, 248
serotonin, 20, 248
side effects, drug and treatment, 2, 41–42, 44, 47–48, 120, 161, 175, 178, 184, 187, 193–94, 200, 210
signature field, 31
Sills, Franklin
    The Polarity Principle, 94
soul, 1–3, 5–6, 15, 29, 33–37, 74–75, 89, 101, 134, 153, 156–57, 186, 210, 212–14, 217–19
Source, 79, 91–93, 111, 168, 178
spirit, 2–3, 5–6, 15, 29, 33, 41, 43, 45–46, 48, 65, 93–94, 98, 135, 149, 156–57
spirituality, 15, 143, 157, 239, 248
spontaneous healing, 41, 218
Stone, Randolph, 94, 99
stress, 3, 9, 15, 21–25, 104
stroke, 22, 110, 117–18, 168, 177, 180–82, 187
Sullivan, Annie, 35

superconducting quantum interference device (SQUID), 14, 65, 121
superstring theory, 90, 94, 239, 248
surgery, 42–43, 47–48, 142, 154, 179, 187, 194, 205, 222
synapses, 117, 236, 248
Syracuse University, 64, 248
Szent-Györgyi, Albert, 63, 65, 248

T

Tainio, Bruce, 32
Tao ("The Way"), 91, 248
telomeres, 46, 248
Tesla, 14, 61, 248
Thales of Miletus (Greek philosopher), 62, 248
Totipotency, 10, 248
Touch Therapy (TT), 44, 62, 66, 121, 197
transcutaneous electrical nerve stimulation (TENS), 61
tumor, 10, 63, 65, 76, 117, 119, 128–29, 179, 203
tumor hierarchy, 10, 248
21 Grams (movie), 157

U

Universal Life Force, 91, 178, 239–40, 248
Upanishads, 95, 248

V

vata, 42, 81–82, 99, 233, 240, 248
vibration, 31, 34, 58, 63, 77, 79, 87, 91, 121
vibrational frequencies, 32, 61–62, 65, 75, 95, 103, 240, 248
vitamins, 108, 162, 166, 174, 203, 243, 248
Voll, Reinhold, 63

W

Waddington, C. H., 10
Walsch, Neale Donald, 212
    Home with God: In a Life that Never Ends, 211
wave function, 103

What is Life? The Physical Aspect of the Living Cell (Schrödinger), 132
"Wisdom of the Body: Future Techniques and Approaches to Morphogenetic Fields in Regenerative Medicine, Developmental Biology, and Cancer" (Levin), 127
World Health Organization (WHO), 177, 204, 248

Y

Yale University, 63, 249
yin and yang, 92–93, 196
You Have Chosen to Remember: A Journey from Perception to Knowledge, Peace of Mind & Joy (Cisneros), 212
Young, Gary, 32

www.ingramcontent.com/pod-product-compliance
Lightning Source LLC
Chambersburg PA
CBHW031831170526
45157CB00001B/264